Multi-Frame Motion-Compensated Prediction for Video Transmission

THE KLUWER INTERNATIONAL SERIES IN ENGINEERING AND COMPUTER SCIENCE

MULTI-FRAME MOTION-COMPENSATED PREDICTION FOR VIDEO TRANSMISSION

THOMAS WIEGAND
Heinrich Hertz Institute

BERND GIROD
Stanford University

Kluwer Academic Publishers
Boston/Dordrecht/London

Distributors for North, Central and South America:
Kluwer Academic Publishers
101 Philip Drive
Assinippi Park
Norwell, Massachusetts 02061 USA
Telephone (781) 871-6600
Fax (781) 681-9045
E-Mail <kluwer@wkap.com>

Distributors for all other countries:
Kluwer Academic Publishers Group
Distribution Centre
Post Office Box 322
3300 AH Dordrecht, THE NETHERLANDS
Telephone 31 78 6392 392
Fax 31 78 6546 474
E-Mail <services@wkap.nl>

Electronic Services <http://www.wkap.nl>

Library of Congress Cataloging-in-Publication Data

Wiegand, Thomas.
Multi-frame motion-compensated prediction for video transmission / Thomas Wiegand, Bernd Girod.
p.cm.—(Kluwer international series in engineering and computer science; SECS 636)
Includes bibliographical references and index.
ISBN 0-7923-7497-5 (alk. paper)
1. Digital video. 2. Video compression. 3. Coding theory. I. Girod, Bernd. II. Title. III. Series.

TK6680.5 .W54 2001
621.388—dc21

2001046191

Printed on acid-free paper.

Printed in the United States of America.

Contents

Foreword

This body of work by Thomas Wiegand and Bernd Girod has already proved to have an exceptional degree of influence in the video technology community, and I have personally been in a position to proudly witness much of that influence.

I have been participating heavily in the video coding standardization community for some years – recently as the primary chairman ("rapporteur") of the video coding work in both of the major organizations in that area (the ITU-T VCEG and ISO/IEC MPEG organizations). The supporters of such efforts look for meritorious research ideas that can move smoothly from step to step in the process found there:

- generation of strong proposal descriptions,
- tests of effectiveness,
- adjustments for practicality and general flexibility, and
- precise description in a final approved design specification.

The ultimate hope in the standardization community is that the specifications written there and the other contributions developed there will prove to provide all the benefits of the best such efforts:

- enabling the growth of markets for products that work well together,
- maximizing the quality of these products in widespread use, and
- progressing the technical understanding of the general community.

The most well-known example of such a successful effort in the video coding community is the MPEG-2 video standard (formally identified as ITU-T Recommendation H.262 or as ISO/IEC International Standard 13818-2). MPEG-2 video is now used for DVD Video disks, direct-broadcast satellite services, terrestrial broadcast television for conventional and high-definition services, digital cable television, and more. The MPEG-2 story owes some of its success to lessons learned in earlier standardization efforts – including the first digital video coding standard known as ITU-T Recommendation H.120, the first truly practical success known as ITU-T Recommendation H.261 (a standard that enabled the growth of the new industry of videoconferencing), and the MPEG-1 video standard (formally ISO/IEC 11172-2, which enabled the storage of movies onto inexpensive compact disks). Each generation of technology has benefited from lessons learned in previous efforts.

The next generation of video coding standard after MPEG-2 is represented by ITU-T Recommendation H.263 (a standard primarily used today for videoconferencing, although showing strong potential for use in a variety of other applications), and it was the "H.263++" project for enhancing that standard that provided a key forum for Wiegand and Girod's work.

At the end of 1997, Thomas Wiegand, Xiaozheng Zhang, Bernd Girod, and Barry Andrews brought a fateful contribution (contribution Q15-C-11) to the Eibsee, Germany meeting of the ITU-T Video Coding Experts Group (VCEG). In it they proposed their design for using long-term memory motion-compensated prediction to improve the fidelity of compressed digital video. The use of long-term memory had already begun to appear in video coding with the recent adoption of the error/loss resilience feature known as reference picture selection or as "NEWPRED" (adopted into H.263 Annex N with final approval in January of 1998 and also adopted about two years later into the most recent ISO/IEC video standard, MPEG-4). But the demonstration of a way to use long-term memory as an effective means of improving coded video quality for reliable channels was clearly new and exciting.

Part of the analysis in that contribution was a discussion of the importance of using good rate-distortion optimization techniques in any video encoding process. The authors pointed out that the reference encoding method then in use by VCEG (called the group's *test model number 8*) could be significantly improved by incorporating better rate-distortion optimization. It was highly admirable that, in the interest of fairness, part of the proposal contribution was a description of a method to improve the quality of the reference *competition* against which their proposal would be evaluated. It was in this contribution that I first saw the simple equation

$$\lambda_{\mathrm{MOTION}} = \sqrt{\lambda_{\mathrm{MODE}}}\,.$$

A few months later (in VCEG contribution Q15-D-13), Wiegand and Andrews followed up with the extremely elegant simplification

$$\lambda_{\mathrm{MODE}} = 0.85 \cdot Q^2\,.$$

For years (starting with the publication of a paper by Yair Shoham and Allen Gersho in 1988), the principles of rate-distortion optimization had become an increasingly-familiar concept in the video compression community. Many members of the community (myself included, starting in 1991) had published work on the topic – work that was all governed by a frustrating little parameter known as λ. But figuring out what value to use for λ had long been a serious annoyance. It took keen insight and strong analysis to sort out the proper relationship between a good choice for λ and Q, the parameter governing the coarseness of the quantization. Wiegand, working under the tutelage of Girod and in collaboration with others at the University of Erlangen-Nuremberg and at 8x8, Inc. (now Netergy Networks), demonstrated that insight and analytical strength.

The ITU-T VCEG adopted the rate-distortion optimization method into its test model immediately (in April of 1998), and has used that method ever since. It is now preparing to adopt a description of it as an appendix to the H.263 standard to aid those interested in using the standard. I personally liked the technique so much that I persuaded Thomas Wiegand to co-author a paper with me for the November, 1998 issue of the IEEE Signal Processing Magazine and include a description of the method. And at the time of this writing, the ISO/IEC Moving Picture Experts Group (MPEG) is preparing to to conduct some tests against a reference level of quality produced by its recent MPEG-4 video standard (ISO/IEC International Standard 14496-2) – and it appears very likely that MPEG will also join the movement by choosing a reference that operates using that same rate-distortion optimization method.

But long-term memory motion compensation was the real subject of that 1997 contribution, while the rate-distortion optimization was only a side note. The main topic has fared even better than the aside. The initial reaction in the community was not one of unanimous enthusiasm – in fact some thought that the idea of increasing the memory and search requirements of video encoders and decoders was highly ill-advised. But diligence, strong demonstrations of results, and perhaps more iteration of Moore's Law soon persuaded the ITU-T VCEG to adopt the long-term memory feature as Annex U to Recommendation H.263. After good cross-verified *core experiment* results were shown in February of 1999, the proposal was adopted as draft Annex U. Additional good work described in this text in regard to fast search methods helped in convincing the skeptics of the practicality of using long-term memory. Ultimately, draft Annex U was adopted as a work item and evolved to preliminary approval in February of 2000 and then final approval in November of 2000.

A remarkable event took place in Osaka in May of 2000, when Michael Horowitz of Polycom, Inc. demonstrated an actual real-time implementation of Annex U in a prototype of a full videoconferencing product (VCEG contribution Q15-J-11). Real-time efficacy demonstrations of preliminary draft video coding standards has been an exceedingly rare thing in recent years. The obvious improvement in quality that was demonstrated by Horowitz's system was sufficient to squelch even the smallest grumblings of criticism over the relatively small cost increases for memory capacity and processing power.

In only three years, the long-term memory proposal that started as a new idea in a university research lab has moved all the way to an approved international standard and real market-ready products with obvious performance benefits. That is the sort of rapid success that research engineers and standards chairmen dream about at night.

Even newer ways of using long-term memory (such as some error resilience purposes also described in this work) have begun to appear and mature. Other concepts described in this work (such as affine multi-frame motion compensation) may one day also be seen as the initial forays into the designs for a new future.

As the community has grown to appreciate the long-term memory feature, it has become an embraced part of the conventional wisdom. When the ITU-T launched an initial design in August of 1999 for a next-generation "H.26L" video coding algorithm beyond the capabilities of today's standards, Wiegand's long-term memory idea was in it from the very beginning. The tide has turned. What once seemed like the strange and wasteful idea of requiring storage and searching of extra old pictures is becoming the accepted practice – indeed it is the previous practice of throwing away the old decoded pictures that has started to seem wasteful.

GARY J. SULLIVAN, PH.D.
Rapporteur of ITU-T VCEG (ITU-T Q.6/SG16 Video Coding Experts Group),
Rapporteur of ISO/IEC MPEG Video (ISO/IEC JTC1/SC29/WG11
Moving Picture Experts Group Video Subgroup),
Microsoft Corporation Software Design Engineer

Preface

In 1965, Gordon Moore, when preparing a speech, made a famous observation. When he started to graph data about the growth in memory chip performance, he realized that each new chip had twice as much capacity as its predecessor, and that each chip was released within 18-24 months of the previous chip. This is but one example of exponential growth curves that permeate semiconductor technology and computing. Moore's Law has become synonymous with this exponential growth, but it is nice to remember that memory chips were its first domain.

This book is the result of the doctoral research by one of us (T.W.) under the guidance of the other (B.G.), both working at the time at the Telecommunications Laboratory of the University of Erlangen-Nuremberg, Germany. In 1995, when this very fruitful collaboration started, video compression, after 2 decades of work by many very talented scientists and engineers, seemed very mature. Nevertheless, we were looking for novel ways to push video compression algorithms to even lower bit-rates, while maintaining an acceptable image quality. And we turned to Moore's Law for that.

Thirty years after the formulation of Moore's Law, memory capacity had increased such we could easily store dozens or even hundreds of uncompressed video frames in a single memory chip. We could already foresee the time when a single chip would hold thousands of uncompressed frames. Still, our compression algorithms at the time would only make reference to one previous frame (or maybe 2, as for B-pictures). The question how much better one could do by using many frames had never really been addressed, and we found it intriguing in its simplicity. As so often, the first experimental results were not very encouraging, but financial support by German Science Foundation, combined with the insight that, at least, we should not do worse than with a single-frame technique, kept us going.

In hindsight, the project is a rare example of university research with immediate impact, drawing a straight path from idea to fundamental research to international standardization to commercial products. After an initial phase of investigation, most of the research is this book has been conducted in connection with the ITU-T/SG 16/VCEG standardization projects H.263++ and H.26L. As a result, large parts of the techniques presented in this book have been adopted by the ITU-T/SG 16 into H.263++ and are integral parts of ongoing H.26L project. To our great delight, the first real-time demonstration of our multi-frame prediction technique in a commercial video conferencing system was shown even before the H.263++ standard was finalized. Today, multi-frame motion-compensated prediction appears such a natural component of the video compression tool-box, and we expect to see it being used universally in the future.

This work would not have been possible without the stimulating collaboration and the generous exchange of ideas at the Telecommunications Laboratory at the University of Erlangen-Nuremberg. The authors gratefully acknowledge the many contributions of these former or current members of the Image Communication Group: Peter Eisert, Joachim Eggers, Niko Färber, Markus Flierl, Eckehard Steinbach, Klaus Stuhlmüller, and Xiaozheng Zhang. Moreover, Barry Andrews and Paul Ning at 8x8, Inc. (now Netergy Networks, Inc.) and, last but not least, Gary Sullivan, the Rapporteur of the ITU-T Video Coding Experts Group, are acknowledged for their help and support.

THOMAS WIEGAND AND BERND GIROD

To our families.

Introduction

It has been customary in the past to transmit successive complete images of the transmitted picture. This method of picture transmission requires a band of frequencies dependent on the number of images transmitted per second. Since only a limited band of frequencies is available for picture transmission, the fineness in the detail of the transmitted picture has therefore been determined by the number of picture elements and the number of pictures transmitted per second. In accordance with the invention, this difficulty is avoided by transmitting only the difference between the successive images of an object.

RAY DAVIS KELL
— *Improvements relating to Electric Picture Transmission Systems*
— British Patent, 1929

Video compression algorithms are a key component for the transmission of motion video. The necessity for video compression arises from the discrepancy of the bit-rates between the raw video signal and the available transmission channels. The motion video signal essentially consists of a time-ordered sequence of pictures, typically sampled at 25 or 30 pictures per second. Assume that each picture of a video sequence has a relatively low Quarter Common Intermediate Format (QCIF) resolution, i.e., 176×144 samples, that each sample is digitally represented with 8 bits, and that two out of every three pictures are skipped in order to cut down the bit-rate. For color pictures, three color component samples are necessary to represent a sufficient color space. In order to transmit even this relatively low-resolution sequence of pictures, the raw video bit-rate is still more than 6 Mbit/s.

On the other hand, today's low-cost transmission channels for personal communications often operate at much lower bit-rates. For instance, V.34 modems transmit at most 33.4 kbit/s over dial-up analog phone lines. Although, the digital subscriber loop [Che99] and optical fiber technology are rapidly advancing, bit-rates below 100 kbit/s are typical for most Internet connections today. For wireless transmission, bit-rates suitable for motion video can be found only to a very limited extent. Second-generation wireless networks, such as Global System for Mobile Communications (GSM), typically provide 10–15 kbit/s which is too little for motion video. Only the Digital Enhanced Cordless Telecommunications (DECT) standard with its limited local support can be employed providing bit-rates of 32, 80, or more kbit/s [PGH95]. Third-generation wireless networks are well underway and will provide increased bit-rates [BGM+98]. Nevertheless, bit-rate remains to be a valuable resource and therefore, the efficient transmission of motion video will be important in the future. One way towards better video transmission systems is to increase the efficiency of the video compression scheme, which is the main subject of this book. Furthermore, the robustness of the system in case of transmission errors is an important issue which is considered in this book as well.

In the early 1980s, video compression made the leap from intra-frame to inter-frame algorithms. Significantly lower bit-rates were achieved by exploiting the statistical dependencies between pictures at the expense of memory and computational requirements that were two orders of magnitude larger. Today, with continuously dropping costs of semiconductors, one might soon be able to afford another leap by dramatically increasing the memory in video codecs to possibly hundreds or even thousands of reference frames. Algorithms taking advantage of such large memory capacities, however, are in their infancy today. This has been the motivation for the investigations into multi-frame motion-compensating prediction in this book.

I.1 MAIN CONTRIBUTIONS

In most existing video codecs today, inter-frame dependencies are exploited via motion-compensated prediction (MCP) of the original frame by referencing the prior decoded frame only. This single-frame approach follows the argument that the changes between successive frames are rather small and thus the consideration of short-term statistical dependencies is sufficient. In this book it is demonstrated that long-term statistical dependencies can be successfully exploited with the presented approach: multi-frame MCP. The main contributions of this book are as follows:

- It is demonstrated that the combination of multi-frame MCP with Lagrangian bit-allocation significantly improves the rate-distortion performance of hybrid video coding. For multi-frame prediction, motion compensation is extended from referencing the prior decoded frame to several frames. For that, the motion vector utilized in block-based motion compensation is extended by a picture reference parameter.

- An efficient approach to Lagrangian bit-allocation in hybrid video coding is developed. The concepts of rate-constrained motion estimation and coding mode decision are combined into an efficient control scheme for a video coder that is based on ITU-T Recommendation H.263. Moreover, a new approach for choosing the coder control parameter is presented and its efficiency is demonstrated. The comparison to a previously known bit-allocation strategy shows that a bit-rate reduction up to 10 % can be achieved using the H.263-based anchor that uses Lagrangian bit-allocation.

- Long-term memory MCP is investigated as a means to exploit long-term statistical dependencies in video sequences. For long-term memory MCP, multiple past decoded pictures are referenced for motion compensation. A statistical model for the prediction gain is developed that provides the insight that the PSNR improvements in dB are roughly proportional to the log-log of the number of reference frames.

- Long-term memory MCP is successfully integrated into an H.263-based hybrid video codec. For that, the Lagrangian bit allocation scheme is extended to long-term memory MCP. Experimental results are presented that validate the effectiveness of long-term memory MCP. Average bit-rate savings of 12 % against the H.263-based anchor are obtained, when considering 34 dB reproduction quality and employing 10 reference frames. When employing 50 reference frames, the average bit-rate savings against the H.263-based anchor are 17 %. For some image sequences, very significant bit-rate savings of more than 60 % can be achieved.

- The concept of long-term memory MCP is taken further by extending the multi-frame buffer with warped versions of decoded frames. Affine motion parameters describe the warping. A novel coder control is proposed, that determines an efficient number of affine motion parameters and reference frames. Experimental results are presented that demonstrate the efficiency of the new approach. When warping the prior decoded frame, average bit-rate savings of 15 % against the H.263-based anchor are reported for the case that 20 additional reference pictures are warped. Further experiments show that the combination of long-term memory MCP and reference picture warping provides almost additive rate-distortion gains. When employing 10 decoded reference frames and 20 warped reference pictures, average bit-rate savings of 24 % against the H.263-based anchor can be obtained. In some cases, the combination of affine and long-term memory MCP provides more than additive gains.

- Novel techniques for fast multi-frame motion estimation are presented, which show that the computational requirements can be reduced by more than an order of magnitude, while maintaining all or most of the improvements in coding efficiency. The main idea investigated is to pre-compute data about the search space of multiple reference frames that can be used to either avoid considering certain positions or to reduce the complexity for evaluating distortion. The presented results indicate that the increased computational complexity for multi-frame motion estimation is not an obstacle to practical systems.

- The efficiency of long-term memory MCP is investigated for channels that show random burst errors. A novel approach to coder control is proposed incorporating an estimate of the average divergence between coder and decoder given the statistics of the random channel and the inter-frame error propagation. Experimental results incorporating a wireless channel show, that long-term memory MCP significantly outperforms the H.263-based anchor in the presence of error-prone channels for transmission scenarios with and without feedback.

I.2 PRACTICAL IMPORTANCE

Practical communication is impossible without specifying the interpretation of the transmitted bits. A video coding standard is such a specification and most of today's practical video transmission systems are standard compliant. In recent years, the ITU-T Video Coding Experts Group has been working on the ITU-T/SG16/Q.15 project which resulted in the production of the very popular H.263 video coding standard.

H.263, version 1, was approved in early 1996 by the ITU-T with technical content completed in 1995. H.263 was the first codec designed specifically to handle very low bit-rate video, and its performance in that arena is still state-of-the-art [ITU96a, Rij96, GSF97]. But, H.263 has emerged as a high compression standard for moving images, not exclusively focusing on very low bit-rate applications. Its original target bit-rate range was about 10-30 kbit/s, but this was broadened during development to perhaps 10-2048 kbit/s. H.263, version 2, was approved in January of 1998 by the ITU-T with technical content completed in September 1997 [ITU98a]. It extends the effective bit-rate range of H.263 to essentially any bit-rate and any progressive-scan (non-interlace) picture format. Some ideas that are described in this book have been successfully proposed to the ITU-T Video Coding Experts Group as technical contributions to H.263, version 3, and the succeeding standardization project H.26L. The following achievements have been made:

- The proposal for a Lagrangian coder control [ITU98b] including the specifications for the parameter settings lead to the creation of a new encoder test model, TMN-10, for the ITU-T Recommendation H.263, version 2. The encoder test model is an informative recommendation of the ITU-T Video Coding Experts Group for the H.263 video encoder. Further, the approach to Lagrangian coder control has also been adopted for the test model of the new standardization project of the ITU-T Video Coding Experts Group, H.26L.

- The long-term memory MCP scheme has been accepted as an Annex of ITU-T Recommendation H.263, version 3 [ITU99b]. The currently ongoing project of the ITU-T Video Coding Experts Group, H.26L, incorporates long-term memory MCP from the very beginning as an integral part.

I.3 ORGANIZATION OF THE BOOK

The combination of multi-frame MCP with Lagrangian bit-allocation is an innovative step in the field of video coding. Once this step was taken, a large variety of new opportunities and problems appeared. Hence, this book addresses the variety of effects of multi-frame MCP which are relevant to bit-allocation, coding efficiency, computational complexity, and transmission over error-prone channels. This book is organized as follows:

In Chapter 1, "State-of-the-Art Video Transmission", the considered transmission framework and today's most successful approaches to source coding of motion video are presented. The features of the H.263 video coding standard are explained in detail. H.263 is widely considered as state-of-the-art in video coding and is therefore used as the underlying framework for the evaluation of the ideas in this book.

In Chapter 2, "Rate-Constrained Coder Control", the operational control of the video encoder is explained. Attention is given to Lagrangian bit-allocation which has emerged as a widely accepted approach to efficient coder control. TMN-10, which is the recommended coder control for ITU-T Recommendation H.263 is explained in detail since parts of it have been developed in this book. Moreover, TMN-10 serves as the underlying bit-allocation scheme for the various new video coding approaches that are being investigated in Chapters 3-6.

In Chapter 3, "Long-Term Memory Motion-Compensated Prediction", the multi-frame concept is explained with a particular emphasis on long-term memory MCP, the scheme adopted in Annex U of H.263++ [ITU00]. The implications of the multi-frame approach on the video syntax and bit-allocation are investigated. The dependencies that are exploited by long-term memory MCP are analyzed and statistically modeled. Experimental results verify the coding efficiency of long-term memory MCP.

In Chapter 4, "Affine Multi-Frame Motion-Compensated Prediction", the extension of the translational motion model in long-term memory MCP to affine motion models is explained. An extension of the TMN-10 bit-allocation strategy is presented that robustly adapts the number of affine motion parameters to the scene statistics which results in superior rate-distortion performance as verified by experiments.

Chapter 5, "Fast Motion Estimation for Multi-Frame Prediction", presents techniques to reduce the computational complexity that is associated with motion estimation on multiple frames. The focus is on the block matching process in multi-frame MCP. Experiments are presented that illustrate the trade-off between rate-distortion performance and computation time.

In Chapter 6, “Error Resilient Video Transmission”, it is demonstrated that long-term memory MCP can also be successfully applied to improve the rate-distortion performance of video transmission systems in the presence of channel errors. A new coder control is presented that takes into account the decoding distortion including the random transmission errors. Experimental results verify the rate-distortion performance of the new approach for a transmission over a wireless channel with burst errors.

Chapter 1

STATE-OF-THE-ART VIDEO TRANSMISSION

This book discusses ideas to improve video transmission systems via enhancing the rate-distortion efficiency of the video compression scheme. The rate-distortion efficiency of today's video compression designs is based on a sophisticated interaction between various motion representation possibilities, waveform coding of differences, and waveform coding of various refreshed regions. Modern video codecs achieve good compression results by efficiently combining the various technical features. The most successful and widely used designs today are called hybrid video codecs. The naming of these codecs is due to their construction as a hybrid of MCP and picture coding techniques. The ITU-T Recommendation H.263 is an example for a hybrid video codec specifying a highly optimized video syntax.

This chapter is organized as follows. In Section 1.1, the considered video transmission scenario is outlined. In Section 1.2, the basic components of today's video codecs are reviewed with an emphasis on MCP in hybrid video coding, since this book mainly focuses on the MCP part. This section also introduces notation and relevant terms. The ITU-T Recommendation H.263 is an example for an efficient and widely used motion-compensating hybrid video codec and the main features of H.263 are described in Section 1.3. A software realization of H.263 serves as a basis for comparison throughout this book. In Section 1.4, the effectiveness of the motion compensation features in H.263 is presented by means of experimental results. Advanced techniques for MCP that relate to the ideas in this book are reviewed in Section 1.5. Finally, known video source coding techniques that improve the transmission of coded video over error-prone channels are presented in Section 1.6.

1.1 VIDEO TRANSMISSION SYSTEM

An example for a typical video transmission scenario that is considered in this book is shown in Fig. 1.1. The *video capture* generates a space- and time-

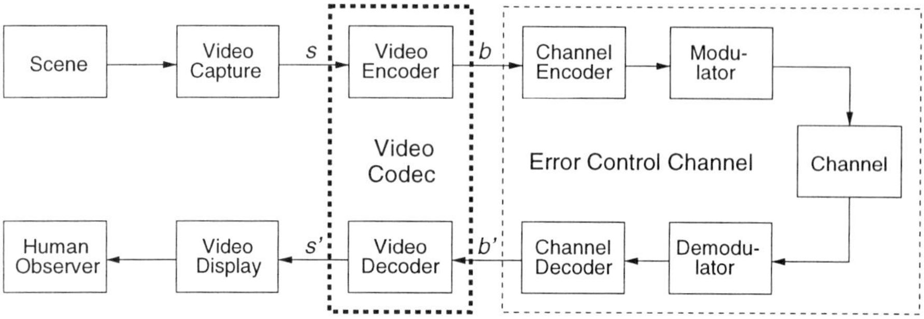

Figure 1.1. Video transmission system.

discrete video signal s, for example using a camera that projects the 3-D *scene* onto the image plane. Cameras typically generate 25 or 30 frames per second and in this book it is assumed that the video signal s is a progressive-scan picture in Common Intermediate Format (CIF) or QCIF resolution. The *video encoder* maps the video signal s into the bit-stream b. The bit-stream is transmitted over the *error control channel* and the received bit-stream b' is processed by the *video decoder* that reconstructs the decoded video signal $\acute{s}$ and presents it via the *video display* to the *human observer*. The quality of the decoded video signal $\acute{s}$ as perceived by the *human observer* is quantified using objective distortion measures. This book focuses on the video encoder and video decoder part with the aim of improved rate-distortion performance of video transmission systems.

The error characteristic of the digital channel can be controlled by the *channel encoder* which adds redundancy to the bits at the video encoder output b. The *modulator* maps the channel encoder output to an analog signal which is suitable for transmission over a physical *channel*. The *demodulator* interprets the received analog signal as a digital signal which is fed into the *channel decoder*. The channel decoder processes the digital signal and produces the received bit-stream b' which may be identical to b even in the presence of channel noise. The sequence of the five components, channel encoder, modulator, channel, demodulator, and channel decoder, are lumped into one box which is called the error control channel. In this book, video transmission systems with and without noisy error control channels, i.e., with and without difference between b and b', are considered.

Common to most transmission scenarios is that there is a trade-off between bit-rate, transmission error rate, and delay. Each of these quantities affects video compression and transmission to a large extent. The bit-rate available to the video encoder controls the distortion and an unreliable channel may cause additional distortion at the decoder. Hence, reducing the bit-rate of the video coder and using the remaining bits for channel coding might improve the overall transmission performance. But the decoder distortions are influenced by a large variety of internal parameters that affect the video syntax, the video decoder, and the coder control. One important external parameter is delay since it is limited in many applications. But increasing the permissible delay can significantly enhance the performance of both, channel and source coding.

This book presents new ideas to enhance the rate-distortion performance of transmission systems via modifications of the video codec, given a limited end-to-end delay found in interactive communication systems. A typical scenario for evaluation of the rate-distortion performance of proposed video coding schemes is as follows. Given video codec A, the anchor, and video codec B, the newly proposed scheme. Evaluate the proposed scheme against the anchor by comparing the quality of the decoded and reconstructed video signal by means of an objective distortion measure given a fixed transmission bit-rate, transmission channel, and delay. The comparison can also be made when fixing distortion and comparing transmission bit-rate. The complexity of video codecs A and B will be stated as additional information rather than employing it as a parameter in the evaluation. The ideas in this book are designated to show performance bounds of video transmission systems that are achieved under well defined conditions. Whether a particular approach should be included into a practical coding system has to be judged considering the available resources for that scenario. The remainder of this chapter and Chapter 2 are designated to the description of the anchor (codec A) that is used for comparison against the new techniques in Chapters 3—6.

1.2 BASIC COMPONENTS OF A VIDEO CODEC

One way of coding a video is simply to compress each picture individually, using an image coding standard such as JPEG [ITU92, PM93] or the still image coding part of H.263 [ITU96a]. The most common "baseline" image coding scheme consists of breaking up the image into equal size blocks of 8×8 pixels. These blocks are transformed by a discrete cosine transform (DCT), and the DCT coefficients are then quantized and transmitted using variable length codes. In the following, this kind of coding scheme is named as INTRA-frame coding, since the picture is coded without referring to other pictures in the video sequence. An important aspect of INTRA coding is its potential to mitigate transmission errors. This feature will be looked at in more detail later.

INTRA-frame coding has a significant drawback which is usually a lower coding efficiency compared to INTER-frame coding for typical video content. In INTER-frame coding, advantage is taken of the large amount of temporal redundancy in video content. Usually, much of the depicted scene is essentially just repeated in picture after picture without any significant change. It should be obvious that the video can be represented more efficiently by coding only the changes in the video content, rather than coding each entire picture repeatedly. This ability to use the temporal domain redundancy to improve coding efficiency is what fundamentally distinguishes video compression from still image compression.

A simple method of improving compression by coding only the changes in a video scene is called conditional replenishment (CR). This term has been coined by MOUNTS in [Mou69]. CR coding was the only temporal redundancy reduction method used in the first digital video coding standard, ITU-T Recommendation H.120 [ITU]. CR coding consists of indicating which areas of a picture can just be repeated, and sending new coded information to replace the changed areas. CR coding thus allows a choice between one of two modes of representation for each image segment, which are called in the following the SKIP mode and the INTRA mode.

However, CR coding has a significant shortcoming, which is its inability to refine an approximation. Often the content of an area of a prior picture can be a good approximation of the new picture, needing only a minor alteration to become a better representation. Hence, frame difference (FD) coding in which a refining frame difference approximation can be sent, results in a further improvement of compression performance.

The concept of FD coding can also be taken a step further, by adding MCP. In the 70s, there has been quite a significant amount of publications that proposed MCP. Often, changes in video content are typically due to the motion of objects in the depicted scene relative to the imaging plane, and a small amount of motion can result in a large difference in the values of the pixels in a picture area, especially near the edges of an object. Hence, displacing an area of the prior picture by a few pixels in spatial location can result in a significant reduction in the amount of information that has to be sent as a frame difference approximation. This use of spatial displacements to form an approximation is known as motion compensation and the encoder's search for the best spatial displacement approximation to use is known as motion estimation. An early contribution which already includes block-matching in the pixel domain which is the method of choice for motion estimation today has been published by JAIN and JAIN in 1981 [JJ81]. The coding of the resulting difference signal for the refinement of the MCP signal is known as displaced frame difference (DFD) coding. Video codecs that employ MCP together with DFD coding are called *hybrid codecs*. Figure 1.2 shows such a hybrid video coder.

Consider a picture of size $w \times h$ in a video sequence, consisting of an array of color component values $(s[\boldsymbol{l}], s_{Cb}[\boldsymbol{l}], s_{Cr}[\boldsymbol{l}])^T$, for each pixel location $\boldsymbol{l} = (x, y, t)^T$, in which x and y are integers such that $0 \leq x < w$ and $0 \leq y < h$. The index t refers to the discrete temporal location of the video frame

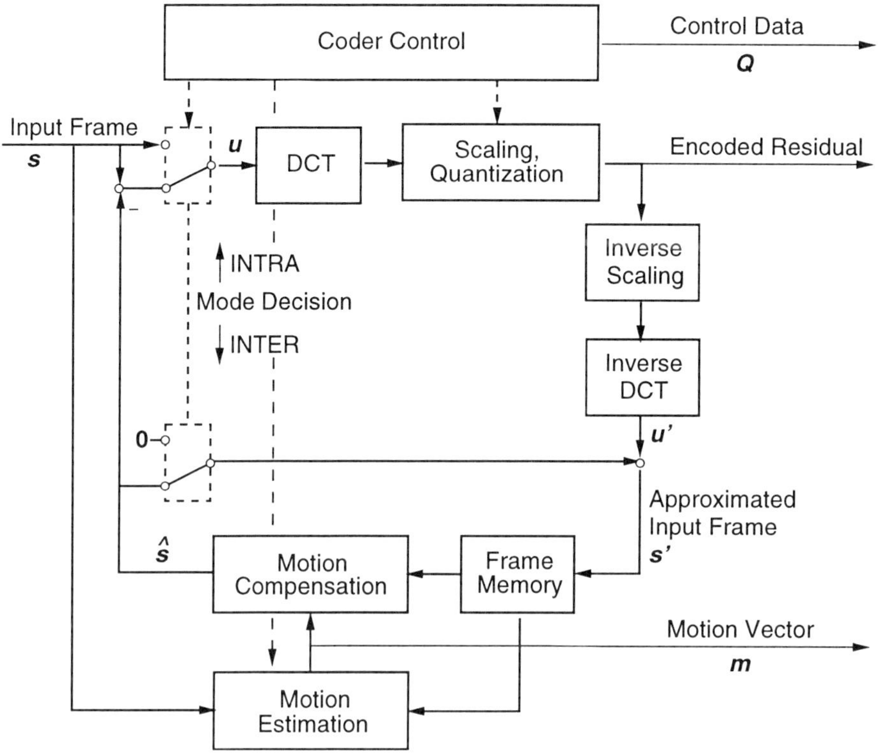

Figure 1.2. A typical hybrid video coder. The space and time discrete input video frame $s[x, y, t]$ and the prior decoded frame $\acute{s}[x, y, t-1]$ are fed into a motion estimation unit. The motion estimation determines the information for the motion compensating predictor. The motion-compensated video frame $\hat{s}[x, y, t]$ is subtracted from the input signal producing the residual video frame $u[x, y, t]$ also called the DFD frame. The residual frame is fed into the residual coder which in many cases consists of a DCT and quantization as well as entropy coding of the DCT coefficients. The approximation of the input video frame $\acute{s}[x, y, t]$ is given as the sum of the motion-compensated frame $\hat{s}[x, y, t]$ and the coded DFD frame $\acute{u}[x, y, t]$. The corresponding hybrid video decoder is run using the control data, the motion vectors and the encoded residual in order to reproduce the same decoded and reconstructed video frame $\acute{s}[x, y, t]$.

and is incremented or decremented by integers of time instants. The decoded approximation of this picture will be denoted as $(\acute{s}[\boldsymbol{l}], \acute{s}_{Cb}[\boldsymbol{l}], \acute{s}_{Cr}[\boldsymbol{l}])^T$. In most video compression systems, the color chrominance components (e.g., $s_{Cb}[\boldsymbol{l}]$ and $s_{Cr}[\boldsymbol{l}]$) are represented with lower resolution (i.e., $\frac{w}{2} \times \frac{h}{2}$) than the luminance component of the image $s[\boldsymbol{l}]$. This is because the human visual system is much more sensitive to brightness than to chrominance, allowing bit-rate savings by coding the chrominance at lower resolution [Wan95]. In such systems, the color chrominance components are motion-compensated using adjusted luminance motion vectors to account for the difference in resolution, since these motion vectors are estimated on the corresponding luminance signals. Hence for the sake of clarity and simplicity, the video signal is in the following regarded as the luminance signal only.

The typical video decoder receives a representation of the current picture which is segmented into K distinct regions $\{\boldsymbol{A}_{k,t}\}_{k=1}^{K}$. For each area, a prediction mode signal $I_k \in \{0,1\}$ is received indicating whether or not the area is predicted. For the areas that are predicted, a motion vector, denoted $\boldsymbol{m}_k = (m_{kx}, m_{ky}, m_{kt})^T$ is received. The motion vector specifies a spatial displacement (m_{kx}, m_{ky}) for motion compensation of that region and the relative reference picture m_{kt} which is usually only the prior decoded picture in standard hybrid video coding. Using the prediction mode and motion vector, a MCP signal $\hat{s}$ is formed for each pixel location $\boldsymbol{l} = (x, y, t)^T$

$$\hat{s}[x,y,t] = I_k \cdot \acute{s}[x - m_{kx}, y - m_{ky}, t - m_{kt}], \quad \text{with} \quad (x,y) \in \boldsymbol{A}_{k,t}\,. \tag{1.1}$$

Please note that the motion vector $\boldsymbol{m}_k$ has no effect if $I_k = 0$ and normally the motion vector is therefore not sent in that case.

In addition to the prediction mode and motion vector information, the decoder receives an approximation $\acute{u}[\boldsymbol{l}]$ of the DFD $u[\boldsymbol{l}]$ between the true image value $s[\boldsymbol{l}]$ and its motion-compensated prediction $\hat{s}[\boldsymbol{l}]$. It then adds the residual signal to the prediction to form the final coded representation

$$\acute{s}[x,y,t] = \hat{s}[x,y,t] + \acute{u}[x,y,t] \quad \text{with} \quad (x,y) \in \boldsymbol{A}_{k,t}. \tag{1.2}$$

Since there is often no movement in large parts of the picture, and since the representation of such regions in the previous picture may be adequate, video coders often utilize the SKIP mode (i.e., $I_k = 1, \boldsymbol{m}_k = (0,0,1)^T, \acute{u}[x,y,t] = 0, \forall (x,y) \in \boldsymbol{A}_{k,t}$) which is efficiently transmitted using very short code words.

In video coders designed primarily for natural scene content, often little freedom is given to the encoder for choosing the segmentation of the picture into regions. Instead, the segmentation is typically either fixed to always consist of a particular two-dimensional block size (typically 16×16 pixels for prediction mode signals and 8×8 for DFD residual content) or in some cases it is allowed to switch adaptively between block sizes (such as allowing the segmentation used for motion compensation to have either a 16×16, 8×8 [ITU96a] or 4×4 [LT00] block size). This is because a pixel-precise segmentation has generally not yet resulted in a significant improvement of compression performance for natural scene content due to the number of bits needed to specify the segmentation, and also because determining an efficient segmentation in an encoder can be a very demanding task. However, in special applications including artificially-constructed picture content rather than natural camera-view scenes, segmented object-based coding may be justified [ISO98b].

1.3 ITU-T RECOMMENDATION H.263

The technical features described above are part of most existing video compression standards including the ISO/IEC JTC 1/SC 29/WG 11 standards MPEG-1 [ISO93], MPEG-2 [ISO94], and MPEG-4 [ISO98b] as well as the ITU-T Recommendations H.261 [ITU93], H.262 (identical to MPEG-2 since it was an official joint project of ISO/IEC JTC 1/SC 29/WG 11 and ITU-T), and H.263 [ITU96a]. The latter, H.263, is described in detail since it is used throughout this book for comparison.

H.263 uses the typical basic structure that has been predominant in all video coding standards since the development of H.261 in 1990, where the image is partitioned into macroblocks of 16×16 luminance pixels and 8×8 chrominance pixels. The coding parameters for the chrominance signals are most of the time inherited from the luminance signals and need only about 10% of the bit-rate. Therefore, the chrominance signals will be ignored in the following and a macroblock is referred to as a block 16×16 luminance pixels.

Each macroblock can either be coded in INTRA or one of several predictive coding modes. In INTRA mode, the macroblock is further divided into blocks of size 8×8 pixels and each of these blocks is coded using DCT, scalar quantization, and run-level variable-length entropy coding. The predictive coding modes can either be of the types SKIP, INTER, or INTER+4V. For the SKIP mode, just one bit is spent to signal that the pixels of the macroblock are repeated from the prior coded frame. The INTER coding mode uses blocks of size 16×16 pixels and the INTER+4V coding mode uses blocks of size 8×8 pixels for motion compensation. For both modes, the MCP error image is encoded similarly to INTRA coding by using the DCT for 8×8 blocks, scalar quantization, and run-level variable-length entropy coding. The motion compensation can be conducted using half-pixel accurate motion vectors where the intermediate positions are obtained via bi-linear interpolation. Additionally, the coder utilizes overlapped block motion compensation, picture-extrapolating motion vectors, and median motion vector prediction.

H.263+ is the second version of H.263 [ITU96a, CEGK98], where several optional features are added to H.263 as Annexes I-T. One notable technical advance over prior standards is that H.263+ was the first video coding standard to offer a high degree of error resilience for wireless or packet-based transport networks. In Section 1.6, source coding features including those that are specified in H.263+ are reviewed that improve rate-distortion performance when transmitting compressed video over error prone channels.

H.263+ also adds some improvements in compression efficiency for INTRA-frame coding. This advanced syntax for INTRA-frame coding is described in Annex I of the ITU-T Recommendation H.263+ [ITU98a]. Annex I provides significant rate-distortion improvements between 1 and 2 dB compared to the H.263 baseline INTRA-frame coding mode when utilizing the same amount of bits for both codecs [CEGK98]. Hence, the advanced INTRA-frame coding scheme of Annex I will be employed in all comparisons throughout this book.

Other Annexes contain additional functionalities including specifications for custom and flexible video formats, scalability, and backward-compatible supplemental enhancement information. The syntax of H.263+ [ITU98a] provides the underlying structure for tests of the MCP ideas in this book.

1.4 EFFECTIVENESS OF MOTION COMPENSATION TECHNIQUES IN HYBRID VIDEO CODING

In Section 1.2, the various technical features of a modern video codec are described and in Section 1.3 their integration into the efficient syntax of the H.263 video compression standard is delineated. In this section, the impact of the various parts are assessed via rate-distortion results that are obtained under the the simulation conditions that are described in Appendix A. The distortion is measured as average PSNR as described in Section A.1, while the set of test sequences is specified in Tab. A.1 in Section A.2.

The set of test sequences has been encoded with different prediction modes enabled. For that, each sequence is encoded in QCIF resolution using the H.263+ video encoder incorporating optimization methods described later in Chapter 2. For comparison, rate-distortion curves have been generated and the bit-rate is measured at equal PSNR of 34 dB. The intermediate points of the rate-distortion curves are interpolated and the bit-rate that corresponds to a given PSNR value is obtained. The percentage in bit-rate savings corresponds to different absolute bit-rate values for the various sequences. Hence, also rate-distortion curves are shown. Nevertheless, computing bit-rate savings might provide a meaningful measure, for example, for video content providers who want to guarantee a certain quality of the reconstructed sequences.

The experiments are conducted so as to evaluate the improvements that are obtained when increasing the capability of motion compensation. Please note that all MCP methods tested are included in ITU-T Recommendation H.263+ and therefore the results presented are obtained by enabling prediction modes that correspond to the various cases. The following cases have been considered:

- **INTRA**: INTRA-frame coding. The advanced INTRA-frame coding mode of H.263+ is employed utilizing 8×8 DCT and transform coefficient prediction within each frame [ITU98a].

- **CR**: Conditional replenishment. CR allows a choice between one of two modes of representation for each 16×16 image block (SKIP mode and INTRA mode). SKIP mode means copying the image content from the previous frame.

- **FD**: Frame difference coding. As CR but frame difference coding is enabled additionally, i.e., the copied macroblock can be refined using DCT-based residual coding as specified in H.263+.

- **IP-MC**: Integer-pixel motion compensation. In addition to the prediction options in FD, also full-pixel accurate motion vectors are used. This is the method for motion compensation as specified in H.261 [ITU93]. IP-MC is realized by not searching half-pixel positions in the H.263+ coder. Please note that no loop filter is utilized in the experiments. Such a loop filter as it is specified in H.261 [ITU93] can provide significant improvements in rate-distortion performance [GSF97].

- **HP-MC**: Half-pixel motion compensation. The accuracy of the motion vectors is increased to half-pixel intervals. This case corresponds to the syntax support for motion compensation of the H.263 baseline coder.

- **TMN-10**: Test-model near-term 10, using the recommended H.263+ coder control [ITU98d]. TMN-10 utilizes all coding options from INTRA-frame coding to half-pixel motion compensation. In H.263+ terminology: the coder uses H.263 baseline and Annexes D, F, I, J, and T. The main additional feature is that the encoder can either choose between blocks of size 16×16 and 8×8 for motion compensation.

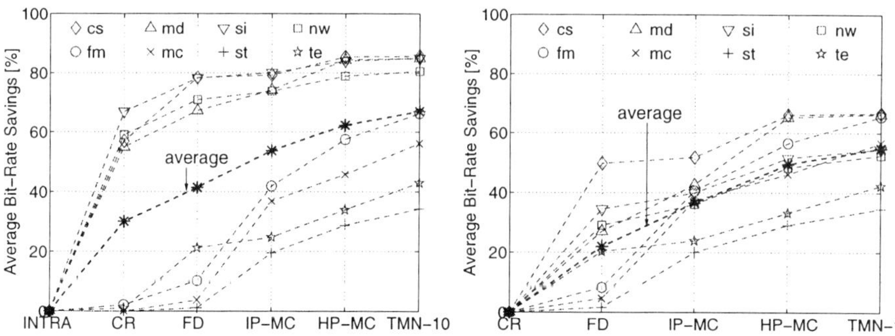

Figure 1.3. Average bit-rate savings versus increased prediction capability for the test sequences in Tab. A.1. The plot on the left-hand side shows the bit-rate savings when setting the bit-rate for advanced intra coding (INTRA) to 100 %. The right-hand side plot shows the same results but setting the bit-rate needed for CR coding to 100 %. The abbreviations fm, mc, st, te, cs, md, nw, and si correspond to those in Tab. A.1 and thus show the bit-rate savings for each test sequence.

Figure 1.3 shows the average reduction in bit-rate for identical PSNR level of 34 dB over the set of test sequences. The left-hand side plot of Fig. 1.3 shows the bit-rate savings when setting the bit-rate for advanced intra coding to 100 %. The bit-rate savings obtained when moving from INTRA-frame to CR coding are dependent on the presence of global motion in the scene. Those 4 sequences for which CR coding provides roughly 60 % bit-rate savings are the ones with a still camera. Obviously, improved motion prediction capability does not provide much additional gain for those areas in the scene that correspond to static background. The other 4 sequences contain global motion and concepts such as CR coding or FD coding do not give a significant improvement against INTRA coding. The average of the 8 sequences is marked with stars.

In order to reduce the influence of the static background, CR coding is used as the reference for determining the bit-rate savings in the right-hand side plot in Fig. 1.3. For that, the bit-rate of CR coding is set to 100 %. A bit-rate saving of 22 % against CR coding can be obtained by FD coding. The next step, full-pixel accurate motion compensation, provides a bit-rate reduction of 15 % on top of FD coding. The step from full-pixel to half-pixel accuracy for 16×16 blocks corresponds to another 13 % of bit-rate savings. This improvement is also theoretically described in [Gir87, Gir93]. The final step, TMN-10, which includes features such as variable block sizes and motion vectors over picture boundaries provides another 5 % bit-rate reduction when considering the bit-rate of CR coding as 100 %. The TMN-10 coder is the anchor that is used throughout this book for comparison.

In Fig. 1.4, the rate-distortion curves for the sequences *Foreman*, *Mobile & Calendar*, *Mother & Daughter*, and *News* from the set of test sequences in Tab. A.1 are shown. For that, the DCT quantization parameter is varied over the values 4, 5, 7, 10, 15, and 25. Other encoding parameters are adjusted accordingly. The precise coder control is following the ideas of TMN-10, the test model of the H.263 standard [ITU98d], which will be explained in the next chapter. The upper two sequences, *Foreman* and *Mobile & Calendar* in Fig. 1.4 contain global motion while the lower ones, *Mother & Daughter* and *News* are captured by a still camera showing only little motion. The gains in PSNR when comparing the cases of CR coding and TMN-10 at equal bit-rates are between 3 and 6 dB.

1.5 ADVANCED MOTION COMPENSATION TECHNIQUES

This section reviews ideas for improving the efficiency of video codecs by further enhancing MCP beyond what is already included in the experiments in the previous section and not covered in this book. Further, only those approaches are reviewed that are related to the ideas that are developed in this book, which are based on:

1. *Exploitation of long-term statistical dependencies,*

2. *Efficient modeling of the motion vector field,*

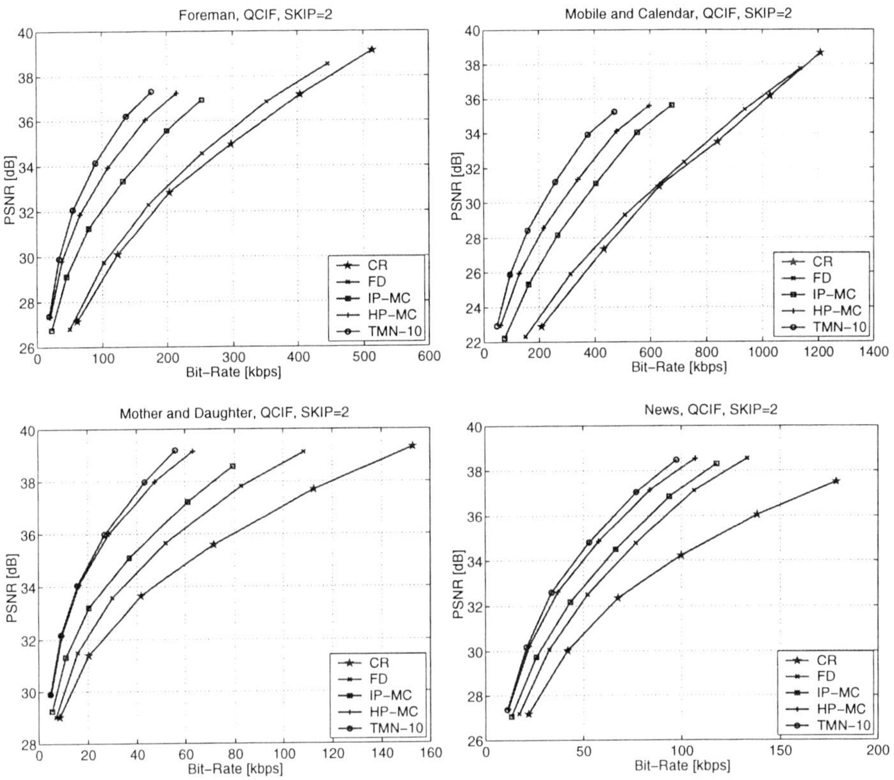

Figure 1.4. Rate-distortion curves for the sequences *Foreman* (top left), *Mobile & Calendar* (top right), *Mother & Daughter* (bottom left), and *News* (bottom right).

3. *Multi-hypothesis prediction.*

For these areas, researchers have developed models and proposed video coding strategies that are described in detail below.

1.5.1 EXPLOITATION OF LONG-TERM STATISTICAL DEPENDENCIES

Long-term statistical dependencies are not exploited in existing video compression standards. Typically, motion compensation is carried out by exclusively referencing the prior decoded frame. This single-frame approach follows the argument that the changes between successive frames are rather small and thus short-term statistical dependencies are sufficient for consideration. However, various techniques have been proposed in the literature for the exploitation of particular long-term dependencies like scene cuts, uncovered background or aliasing-compensated sub-pixel interpolation using multiple past frames.

One approach to exploit particular long-term dependencies is called *short-term frame memory/long-term frame memory* prediction. It has been proposed to the MPEG-4 standardization group [ISO96a]. As specified in [ISO96a], the encoder is enabled to use two frame memories to improve prediction efficiency. The *short-term frame memory* stores the most recently decoded frame, while the *long-term frame memory* stores a frame that has been decoded earlier. In [ISO96a], a refresh rule is specified that is based on a detection of a scene change in order to exploit repeated scene cuts. This approach is included as a special case of the new technique that is presented in Chapter 3 of this book to exploit long-term statistical dependencies.

In [ISO96a], it is also proposed to include frames into the *long-term frame memory* that are generated by so-called *background memory* prediction. Several researchers have proposed algorithms to exploit uncovered background effects using *background memory* prediction [MK85, Hep90, Yua93, DM96, ZK98]. Generating a *background memory* frame as a second reference picture for MCP is mainly an image segmentation problem where an algorithm has to distinguish between moving foreground objects and the background. Most of the *background memory* estimation algorithms work sufficiently well for scenes with stable background but very often break down if camera motion or background changes occur. But the performance of the approach highly depends on the segmentation result. Moreover, the boundary between the foreground and background object has to be composed out of the two frame memories which might increase the bit-rate.

Another approach for improved coding efficiency using long-term dependencies has been presented by WEDI [Wed99]. The approach in [Wed99] employs an advanced sub-pixel motion compensation scheme which is based on the following effect. If an unmoved analog image signal is spatially sampled at the same positions at different times, the two sampled signals are identical, even if the spatial sampling rate is below the Nyquist frequency, in which case the two sampled images would be identical including the resulting aliasing. This also holds, if the image moves by integer factors of the spatial sampling interval. The idea is to assemble a high-resolution image for sub-pixel motion compensation, where the sub-pixel positions are obtained using several past reconstructed frames. Instead of interpolating image content between spatial sampling positions, the corresponding sub-pixel shifted versions in previous decoded frames are utilized. For that, the algorithm in [Wed99] recursively updates the high-resolution image at encoder and decoder simultaneously as the image sequence progresses employing transmitted motion vectors. Again, the performance of this approach highly depends on the estimation step for the high-resolution image.

In Chapter 3, an approach is presented for the exploitation of long-term statistical dependencies called long-term memory MCP. Long-term memory MCP can jointly exploit effects like scene cuts, uncovered background or aliasing-compensated prediction with one single concept. However, long-term memory MCP is not restricted to a particular kind of scene structure or to a particular effect as the above mentioned techniques are.

1.5.2 EFFICIENT MODELING OF THE MOTION VECTOR FIELD

The efficiency of coding the motion information is often increased by enhancing the motion model. This is motivated by the fact that independently moving objects in combination with camera motion and focal length change lead to a sophisticated motion vector field in the image plane which may not be efficiently approximated by purely translational motion models. Also, the exploitation of long-term statistical dependencies might be difficult in this case. Hence, various researchers have proposed techniques to extend the translational motion model towards higher-order parametric models.

In an early work, TSAI and HUANG derive a parametric motion model that relates the motion of planar objects in the scene to the observable motion field in the image plane for a perspective projection model [TH81]. The eight parameters of this model are estimated using corresponding points [TH81]. A problem that very often occurs with the eight parameter model is that some parameters appear in the denominator of the parametric expression which adversely affects the parameter estimation procedure due to numerical problems. In [HT88], HÖTTER and THOMA approximate the planar object motion using a two-dimensional quadratic model of twelve parameters. The parameters are estimated using spatial and temporal intensity gradients which drastically improves the parameter estimates in the presence of noise.

In case the objects in the scene or the considered parts of the objects do not show large depth variations with respect to the image plane, the simpler camera model of parallel projection can be applied. Popular motion models for parallel projection are the affine and bilinear motion model. Various researchers have utilized affine and bilinear motion models for *object-based* or *region-based* coding of image sequences [Die91, San91, YMO95, CAS+96, FVC87, HW98]. The motion parameters are estimated such that they lead to an efficient representation of the motion field inside the corresponding image partition. Due to the mutual dependency of motion estimation and image partition a combined estimation must be utilized. This results in a sophisticated optimization task which usually is very time consuming. Moreover, providing the encoder the freedom to specify a precise segmentation has generally not yet resulted in a significant improvement of compression performance for natural camera-view scene content due to the number of bits needed to specify the segmentation. Hence, other researchers have used affine or bilinear motion models in conjunction with a *block-based* approach to reduce the bit-rate for transmitting the image segmentation [LF95, ZBK97]. They have faced the problem that especially at low bit-rates the overhead associated with higher order motion models that are assigned to smaller size blocks might be prohibitive. A combination of the *block-based* and the *region-based* approach is presented in [KNH97]. KARCZEWICZ *et al.* report in [KNH97] that the use of the twelve parameter motion model in conjunction with a coarse segmentation of the video frame into regions, that consist of a set of connected blocks of size 8×8 pixels, can be beneficial in terms of coding efficiency.

In the previous section, it has been pointed out that *background memory* prediction often breaks down in the case of camera motion. Within the MPEG-4 standardization group, a technique called *Sprites* has been considered [DM96, ISO97b, SSO99] that can be viewed as an extension of *background memory* prediction to robustly handle camera motion. In addition, image content that temporally leaves the field of view can be more efficiently represented. *Sprites* can be used to improve the efficiency of MCP in case of camera motion by warping a second prediction signal towards the actual frame. The technique first identifies background and foreground regions based on local motion estimates. Camera motion is then estimated on the background by applying parametric global motion estimation. After compensating for camera motion, the background content is integrated into a so-called *background mosaic*. The *Sprite* coder warps an appropriate segment of the *background mosaic* towards the current frame to provide the second reference signal. The motion model used is typically a six parameter affine model. The generation of the *background mosaic* is conducted either on-line or off-line and the two approaches are referred to as *Dynamic Sprites* and *Static Sprites*, respectively. So far, only *Static Sprites* are part of the MPEG-4 standard [ISO98a]. For *Static Sprites*, an iterative procedure is applied to analyze the motion in a video sequences of several seconds to arrive at robust segmentation results. This introduces a delay problem that cannot be resolved in interactive applications. On the other hand, the on-line estimation problem for *Dynamic Sprites* is very difficult and only recently some advantages have been reported [SSO99].

An interesting generalization of the *background memory* and *Sprite* techniques has been proposed by WANG and ADELSON, wherein the image sequence is represented by *layers* [WA94]. In addition to the background, the so-called *layered coding* technique can represent other objects in the scene as well. As for *Static Sprites*, the *layers* are determined by an iterative analysis of the motion in a complete image sequence of several seconds.

A simplification of the clustering problem in *object-based* or *region-based* coding and the parameter estimation in *Sprite* and *layered* coding is achieved by restricting the motion compensation to one global model that compensates the camera motion and focal length changes [Höt89, JKS+97, ISO97a]. Often, the background in the scene is assumed to be static and motion of the background in the image plane is considered as camera motion. For the *global motion compensation* of the background often an affine motion model is used where the parameters are estimated typically using two steps. In the first step, the motion parameters are estimated for the entire image and in the second step, the largest motion cluster is extracted. The globally motion-compensated frame is either provided additionally as a second reference frame or the prior decoded frame is replaced. Given the globally motion-compensated image as a reference frame, typically a block-based hybrid video coder conducts translational motion compensation. The drawback of *global motion compensation* is the limitation in rate-distortion performance due to the restriction to one motion parameter vector per frame. The benefits of this approach are the avoidance of sophisticated segmentation and parameter estimation problems. *Global motion compensation* is therefore standardized as an Annex of H.263+ [ITU98a] to enhance the coding efficiency for the on-line encoding of video.

In this book, the *global motion compensation* idea is extended to employing several affine motion parameter sets in Chapter 4. The estimation of the various affine motion parameter sets is conducted so as to handle multiple independently moving objects in combination with camera motion and focal length change. Long-term statistical dependencies are exploited as well by incorporating long-term memory MCP.

1.5.3 MULTI-HYPOTHESIS PREDICTION

Another approach to enhance the performance of motion compensation is multi-hypothesis prediction. The idea of multi-hypothesis MCP is to superimpose various prediction signals to compute the MCP signal. The multi-hypothesis motion-compensated predictor for a pixel location $\boldsymbol{l} = (x, y, t)^T$ in the image segment $\boldsymbol{\mathcal{A}}_k$ is defined as

$$\hat{s}[\boldsymbol{l}] = \sum_{p=1}^{P} h_p[\boldsymbol{l}] \cdot \acute{s}[\boldsymbol{l} - \boldsymbol{m}_{k,p}], \quad \forall \boldsymbol{l} \in \boldsymbol{\mathcal{A}}_k \, . \tag{1.3}$$

with $\hat{s}[\boldsymbol{l}]$ being a predicted pixel value and $\acute{s}[\boldsymbol{l} - \boldsymbol{m}_{k,p}]$ being a motion-compensated pixel from a decoded frame corresponding to the p'th hypothesis. For each of the P hypotheses, the factor h_p specifies the weight that is used to superimpose the various prediction signals. This scheme is a generalization of (1.1) and it includes concepts like sub-pixel accurate MCP [Gir87, Gir93], spatial filtering [ITU93], overlapped block motion compensation (OBMC) [WS91, NO92, Sul93, OS94], and B-frames [MPG85].

The latter approach, B-frames, utilizes two reference frames which are the prior decoded picture and the temporally succeeding picture. In each of the two reference frames, a block is referenced using a motion vector and the MCP signal is obtained by a superposition with identical weights $h_p = 1/2, p = 1, 2$ for both blocks. The weights are constant over the complete block. As the temporally succeeding picture has to be coded and transmitted before the bi-directional predicted picture, a delay problem is introduced that cannot be resolved in interactive applications and therefore, B-frames are not considered further in this book.

A rationale for multi-hypothesis MCP is that if there are P different plausible hypotheses for the motion vector that properly represents the motion of a pixel $s[\boldsymbol{l}]$, and if each of these can be associated with a hypothesis probability $h_p[\boldsymbol{l}]$, then the expected value of the pixel prediction is approximated by (1.3). The expected value is the estimate which minimizes the mean-square error in the prediction of any random variable. Another rationale is that if each hypothesis is viewed as a noisy representation of the pixel, then performing an optimized weighted averaging of the results of several hypotheses as performed in (1.3) can reduce the noise. The multi-hypothesis MCP concept was introduced in [Sul93], and an estimation-theoretic analysis with a focus on OBMC was presented by ORCHARD and SULLIVAN [OS94]. A rate-distortion efficiency analysis including OBMC and B-frames is presented by GIROD in [Gir00].

1.6 VIDEO TRANSMISSION OVER ERROR PRONE CHANNELS

An H.263-compressed video signal is extremely vulnerable to transmission errors. Transmission errors can be reduced by appropriate channel coding techniques. For channels without memory, such as the AWGN channel, channel coding techniques provide very significant reductions of transmission errors at a comparably moderate bit-rate overhead. For the mobile fading channel, however, the effective use of forward error correction is limited when assuming a small end-to-end delay. Here the use of error resilience techniques in the source codec becomes important.

In INTER mode, i.e., when MCP is utilized, the loss of information in one frame has a considerable impact on the quality of the following frames. As a result, spatio-temporal error propagation is a typical transmission error effect for predictive coding. Because errors remain visible for a longer period of time, the resulting artifacts are particularly annoying to end users. To some extent, the impairment caused by transmission errors decays over time due to leakage in the prediction loop. However, the leakage in standardized video decoders like H.263 is not very strong, and quick recovery can only be achieved when image regions are encoded in INTRA mode, i.e., without reference to a previous frame. The INTRA mode, however, is not selected very frequently during normal encoding and completely INTRA coded frames are not usually inserted in real-time encoded video as is done for storage or broadcast applications. Instead, only single macroblocks are encoded in INTRA mode for regions that cannot be predicted efficiently.

The Error Tracking approach [FSG96, SFG97, GF99, FGV98] utilizes the INTRA mode to stop inter-frame error propagation but limits its use to severely impaired image regions only. During error-free transmission, the more effective INTER mode is utilized, and the system therefore adapts to varying channel conditions. Note that this approach requires that the encoder has knowledge of the location and extent of erroneous image regions at the decoder. This can be achieved by utilizing a feedback channel from the receiver to the transmitter. The feedback channel is used to send negative acknowledgment messages (NACKs) back to the encoder. NACKs report the temporal and spatial location of image content that could not be decoded successfully and had to be concealed. Based on the information of a NACK, the encoder can reconstruct the resulting error distribution in the current frame, i.e., track the error from the original occurrence to the current frame. Then, the impaired macroblocks are determined and error propagation can be terminated by INTRA coding these macroblocks.

In this book, the Error Tracking approach is extended to cases when the encoder has no knowledge about the actual occurrence of errors, i.e., without feedback information. In this situation the selection of INTRA coded macroblocks can be done either randomly or preferably in a certain update pattern. For example, ZHU [ZK99b] has investigated update patterns of different shape, such as 9 randomly distributed macroblocks, 1×9, or 3×3 groups of macroblocks. Although the shape of different patterns slightly influences the performance, the selection of the correct INTRA percentage has a significantly higher influence. In [HM92] and [LV96] it is shown that it is advantageous to consider the image content when deciding on the frequency of INTRA coding. For example, image regions that cannot be concealed very well should be refreshed more often, whereas no INTRA coding is necessary for completely static background. In [FSG99, SFLG00], an analytical framework is presented on how to optimize the INTRA refresh rate. In [HPL98], a trellis is used to estimate the concealment quality to introduce a bias into the coder control towards INTRA coding. The extension in this book incorporates an estimate of the spatio temporal error propagation to affect the coder control.

Similar to the Error Tracking approach, the Reference Picture Selection mode of H.263+ also relies upon a feedback channel to efficiently stop error propagation after transmission errors. This mode is described in Annex N of H.263+, and is based on the NEWPRED approach that was suggested in [ITU96b]. A proposal similar to NEWPRED has been submitted to the MPEG-4 standardization group [ISO96b]. Instead of using the INTRA coding of macroblocks, the Reference Picture Selection mode allows the encoder to select one of several previously decoded frames as a reference picture for prediction. In order to stop error propagation while maintaining the best coding efficiency, the available feedback information can be used to select the most recent error-free frame.

Note that also erroneous frames could be used for prediction, if the concealment strategy at the decoder were standardized. In this case, the encoder could exactly reconstruct the erroneous reference frames at the decoder based on NACKs and acknowledgment messages (ACKs). ACKs report the temporal and spatial location of image content that has been decoded successfully. Because of the lack of a standardized concealment strategy and the increase in complexity, this approach is not considered in the description of Annex N. Instead, it is assumed that only error-free frames are selected as a reference. However, for very noisy transmission channels, it can be difficult to transmit complete frames without any errors. In this case, the most recent error-free frame can be very old and hence ineffective for MCP. Therefore, the Independent Segment Decoding mode as described in Annex R of H.263 has been specified. The Independent Segment Decoding mode was suggested in [ITU95]. In the Independent Segment Decoding mode, the video sequence is partitioned into sub-videos that can be decoded independently from each other. A popular choice is to use a group of blocks (GOB) as a sub-video. In a QCIF frame, a GOB consists of a row of 11 macroblocks [ITU98a]. The Independent Segment Decoding mode significantly reduces the coding efficiency of motion compensation, particularly for vertical motion, since image content outside the current GOB must not be used for prediction. Therefore, a simple error concealment strategy is assumed in this book where lost picture content is concealed by the corresponding pixels in the previous decoded picture.

Reference Picture Selection can be operated in two different modes, ACK and NACK mode. In the ACK mode case, correctly received image content is acknowledged and the encoder only uses acknowledged image content as a reference. If the round trip delay is greater than the encoded picture interval, the encoder has to use a reference frame further back in time. This results in decreased coding performance for error-free transmission. In the case of transmission errors, however, only small fluctuations in picture quality occur. In the second mode, the NACK mode, only erroneously received image content is signaled by sending NACKs. During error-free transmission, the operation of the encoder is not altered and the previously decoded image content is used as a reference. Both modes can also be combined to obtain increased performance as demonstrated in [FNI96, TKI97].

BUDAGAVI and GIBSON have proposed multiple reference frames for increased robustness of video codecs [BG96, BG97, BG98]. Error propagation is modeled using a Markov chain analysis which is used to modify the selection of the picture reference parameter using a strategy called *random lag selection*. The Markov chain analysis assumes a simplified binary model of motion compensation not considering quantities like the video signal, actual concealment distortion, the estimation of the spatial displacements and the macroblock mode decision. Hence, the coder control is modified heuristically. In [BG96], also comparisons are presented regarding improved coding efficiency. The comparisons are made against H.263 in baseline mode, i.e., none of the advanced prediction modes that improve coding efficiency was enabled. But a meaningful comparison should include these advanced prediction modes since they significantly change the coding efficiency of H.263-based coding and with that the efficiency trade-off of the components of the transmission system.

The approach that is presented in Chapter 6 of this book unifies concepts such as Error Tracking as well as ACK and NACK reference picture selection into a single approach. For that, an estimate of the average decoder distortion is incorporated into the coder control affecting motion vector estimation and macroblock mode decision.

1.7 CHAPTER SUMMARY

The efficient transmission of video is a challenging task. The state-of-the-art in video coding is represented by the ITU-T Recommendation H.263+. The presented ideas in this book propose to improve the performance of MCP. This approach is in line with past developments in video coding where most of the performance gains have been obtained via enhanced MCP. That is demonstrated by comparing the rate-distortion performance when enabling more and more advanced motion representation possibilities. The best coding efficiency is obtained when all options for motion coding are utilized in H.263+. Hence, H.263+ will be considered as the underlying syntax to evaluate the proposed ideas in this book.

In the literature, various approaches for improving MCP can be found where long-term statistical dependencies in the video sequence are exploited including short-term frame memory/long-term frame memory prediction, background memory prediction, and aliasing prediction. The short-term frame memory/long-term frame memory prediction approach exploits repeated scene cuts. This approach does provide a gain in case such a scene cut occurs and is included as a special case of the new technique that is presented in Chapter 3 of this book. For background memory prediction, researchers have proposed to estimate an additional reference frame for motion compensation that contains the background. For aliasing prediction, a high-resolution image for sub-pixel motion compensation is estimated. The estimation for background memory and aliasing prediction is based on past decoded frames and transmitted parameters since encoder and decoder have to conduct it simultaneously. Therefore, the possible prediction gain highly depends on the accuracy of these estimates. Additionally, each of the methods (short-term frame memory/long-term frame memory, background memory and aliasing prediction) can only exploit the particular effect it is designed for.

Various researchers have proposed to improve the coding efficiency of hybrid video codecs by enhancing the motion model. Typically, affine and bilinear motion models are utilized. In order to provide an efficient representation of the image plane motion, using e.g. affine motion models, the image is often non-uniformly partitioned. Due to the mutual dependency of motion estimation and image partition a combined estimation must be utilized. This results in a sophisticated optimization task which usually is very time consuming. A simplification of the optimization task is achieved by restricting the motion compensation to one global model that compensates the camera motion and focal length changes. The drawback of global motion compensation is the limitation in rate-distortion performance due to the restriction to one motion parameter set per frame. In Chapter 4 of this book, the global motion compensation idea is extended to employing several affine motion parameter sets in Chapter 4.

Another approach for enhancing MCP is called B-frames, where two reference frames are utilized. When coding a block of a B-frame, one block in each of the two reference frames is addressed using a motion vector. The MCP signal is obtained by a superposition with identical weights $1/2$ for both blocks. B-frames can significantly improve prediction performance. However, the two reference frames are the prior decoded picture and the temporally succeeding picture. As the temporally succeeding picture has to be coded and transmitted before the B-frame, a delay problem is introduced that cannot be resolved in interactive applications.

The H.263-compressed video signal is extremely vulnerable to transmission errors. Preventing transmission errors by forward error correction might incur a prohibitive overhead for bursty channels and small end-to-end delays that have to be considered in many applications. Hence, various researchers have proposed video source coding strategies to improve the robustness of video transmission systems. The main problem that is specific to the transmission of hybrid coded video employing MCP is inter-frame error propagation. Known techniques to stop temporal error propagation are INTRA coding and reference picture selection. The application of the new prediction ideas in this book to the transmission of coded video over error-prone channels leads to a generalization of the known techniques with the result of improved rate-distortion performance.

Chapter 2

RATE-CONSTRAINED CODER CONTROL

One key problem in video compression is the operational control of the source encoder. This problem is compounded because typical video sequences contain widely varying content and motion, necessitating the selection between different coding options with varying rate-distortion efficiency for different parts of the image. The task of coder control is to determine a set of coding parameters, and thereby the bit-stream, such that a certain rate-distortion trade-off is achieved for a given decoder. This chapter focuses on coder control algorithms for the case of error-free transmission of the bit-stream. A particular emphasis is on Lagrangian bit-allocation techniques, which have emerged as a widely accepted approach. The popularity of this approach is due to its effectiveness and simplicity.

The application of Lagrangian techniques to control a hybrid video coder is not straightforward because of temporal and spatial dependencies of the rate-distortion costs. The optimization approach presented here concentrates on bit-allocation for the coding parameters for the INTER mode in a hybrid video coding environment. Furthermore, a new and efficient approach to selecting the coder control parameters is presented and evaluated. Based on the coder control developed in this chapter, a contribution was submitted to the ITU-T Video Coding Experts Group [ITU98b], which led to the creation of a new test model, TMN-10 [ITU98d]. TMN-10 is the recommended encoding approach of the ITU-T video compression standard H.263+ [ITU98a]. Moreover, the test model of the new standardization project of the ITU-T Video Coding Experts Group, the TML [LT00], is based on the techniques presented here.

The general approach of bit-allocation using Lagrangian techniques is explained in Section 2.1. Section 2.2 presents a review of known approaches to the application of Lagrangian techniques in hybrid video coding. TMN-10, the encoder test model for the ITU-T Recommendation H.263 is presented in

Section 2.3. In Section 2.4, the approach to choose the parameters for the TMN-10 coder control is described and analyzed by means of experimental results. The efficiency of the proposed techniques is verified by experimental results in Section 2.5. Comparison is made to the threshold-based coder control that is employed in the test model near-term, version 9 (TMN-9). TMN-9 is the preceding ITU-T test model to TMN-10.

2.1 OPTIMIZATION USING LAGRANGIAN TECHNIQUES

Consider K source samples that are collected in the K-tuple $\boldsymbol{\mathcal{S}} = (\boldsymbol{S}_1, \ldots, \boldsymbol{S}_K)$. A source sample $\boldsymbol{S}_k$ can be a scalar or vector. Each source sample $\boldsymbol{S}_k$ can be quantized using several possible coding options that are indicated by an index out of the set $\boldsymbol{\mathcal{O}}_k = \{O_{k1}, \ldots, O_{kN_k}\}$. Let $I_k \in \boldsymbol{\mathcal{O}}_k$ be the selected index to code $\boldsymbol{S}_k$. Then the coding options assigned to the elements in $\boldsymbol{\mathcal{S}}$ are given by the components in the K-tuple $\boldsymbol{\mathcal{I}} = (I_1, \ldots, I_K)$. The problem of finding the combination of coding options that minimizes the distortion for the given sequence of source samples subject to a given rate constraint R_c can be formulated as

$$\begin{aligned} &\min_{\boldsymbol{\mathcal{I}}} D(\boldsymbol{\mathcal{S}}, \boldsymbol{\mathcal{I}}) \\ &\text{subject to} \;\; R(\boldsymbol{\mathcal{S}}, \boldsymbol{\mathcal{I}}) \leq R_c. \end{aligned} \tag{2.1}$$

Here, $D(\boldsymbol{\mathcal{S}}, \boldsymbol{\mathcal{I}})$ and $R(\boldsymbol{\mathcal{S}}, \boldsymbol{\mathcal{I}})$ represent the total distortion and rate, respectively, resulting from the quantization of $\boldsymbol{\mathcal{S}}$ with a particular combination of coding options $\boldsymbol{\mathcal{I}}$. In practice, rather than solving the constrained problem in (2.1), an unconstrained formulation is employed as a Lagrangian minimization approach

$$\boldsymbol{\mathcal{I}}^* = \underset{\boldsymbol{\mathcal{I}}}{\operatorname{argmin}} \left\{ D(\boldsymbol{\mathcal{S}}, \boldsymbol{\mathcal{I}}) + \lambda \cdot R(\boldsymbol{\mathcal{S}}, \boldsymbol{\mathcal{I}}) \right\}, \tag{2.2}$$

with $\lambda \geq 0$ being the Lagrange parameter. This unconstrained solution to a discrete optimization problem was introduced by EVERETT [Eve63]. The solution $\boldsymbol{\mathcal{I}}^*$ to (2.2) is optimal in the sense that if a rate constraint R_c corresponds to λ, then the total distortion $D(\boldsymbol{\mathcal{S}}, \boldsymbol{\mathcal{I}}^*)$ is minimum for all combinations of coding options with bit-rate less or equal to R_c.

Assuming additive distortion and rate measures, the Lagrangian cost function J for a given value of the Lagrange parameter λ can be decomposed into a sum of terms over the elements in $\boldsymbol{\mathcal{S}}$ yielding

$$\boldsymbol{\mathcal{I}}^* = \underset{\boldsymbol{\mathcal{I}}}{\operatorname{argmin}} \sum_{k=1}^{K} J(\boldsymbol{S}_k, \boldsymbol{\mathcal{I}} | \lambda) \tag{2.3}$$

with

$$J(\boldsymbol{S}_k, \boldsymbol{\mathcal{I}} | \lambda) = D(\boldsymbol{S}_k, \boldsymbol{\mathcal{I}}) + \lambda \cdot R(\boldsymbol{S}_k, \boldsymbol{\mathcal{I}}), \tag{2.4}$$

where $D(\boldsymbol{S}_k, \mathcal{I})$ and $R(\boldsymbol{S}_k, \mathcal{I})$ are distortion and rate, respectively, for $\boldsymbol{S}_k$ given the combination of coding options in $\mathcal{I}$. Even with this simplified Lagrangian formulation, the solution to (2.3) remains rather unwieldy due to the rate and distortion dependencies manifested in the $D(\boldsymbol{S}_k, \mathcal{I})$ and $R(\boldsymbol{S}_k, \mathcal{I})$ terms. Without further assumptions, the resulting distortion and rate associated with a particular source sample $\boldsymbol{S}_k$ is inextricably coupled to the chosen coding options for every other source sample in $\mathcal{S}$.

On the other hand, for many coding systems, the bit-stream syntax imposes additional constraints that can further simplify the optimization problem. A computationally very efficient case is obtained when the codec is restricted so that rate and distortion for a given source sample are independent of the chosen coding options of all other source samples in $\mathcal{S}$. As a result, a simplified Lagrangian cost function can be computed as

$$J(\boldsymbol{S}_k, \mathcal{I}|\lambda) = J(\boldsymbol{S}_k, I_k|\lambda). \tag{2.5}$$

In this case, the optimization problem of (2.3) reduces to

$$\min_{\mathcal{I}} \sum_{i=1}^{K} J(\boldsymbol{S}_k, \mathcal{I}|\lambda) = \sum_{i=1}^{K} \min_{I_k} J(\boldsymbol{S}_k, I_k|\lambda), \tag{2.6}$$

and can be easily solved by independently selecting the coding option for each $\boldsymbol{S}_k \in \mathcal{S}$. For this particular scenario, the problem formulation is equivalent to the bit-allocation problem for an arbitrary set of quantizers, proposed by SHOHAM and GERSHO [SG88].

This technique has gained importance due to its effectiveness, conceptual simplicity, and its ability to effectively evaluate a large number of possible coding choices in an optimized fashion. In the next section, the application of Lagrangian optimization techniques to hybrid video coding is described.

2.2 LAGRANGIAN OPTIMIZATION IN VIDEO CODING

Consider a block based hybrid video codec such as H.261, H.263 or MPEG-1/2/4. Let the image sequence s be partitioned into K distinct blocks $\mathcal{A}_k$ and the associated pixels be given as $\boldsymbol{S}_k$. The options $\mathcal{O}_k$ to encode each block $\boldsymbol{S}_k$ are INTRA and INTER coding modes with associated parameters. The parameters are DCT coefficients and quantizer value Q for both modes plus one or more motion vectors for the INTER mode. The parameters for both modes are often predicted using transmitted parameters of preceding modes inside the image. Moreover, the INTER mode introduces a temporal dependency because reference is made to prior decoded pictures via MCP. Hence, the optimization of a hybrid video encoder would require the minimization of the Lagrangian cost function in (2.2) for all blocks in the entire sequence. This minimization would have to proceed over the product space of the coding mode parameters.

Some of the dependencies of the parameters can be represented by a trellis and have indeed been exploited by various researchers using dynamic programming methods. Bit-allocation to DCT coefficients was proposed by ORTEGA and RAMCHANDRAN [OR95], and a version that handles the more complex structure of the entropy coding of H.263 has recently appeared [ITU98c, WLV98]. In [WLCM95, WLM+96], the prediction of coding mode parameters from parameters of preceding blocks inside an image is considered. Interactions such as the number of bits needed to specify a motion vector value that depend on the values of the motion vectors in neighboring regions or the areas of influence of different motion vectors due to overlapped-block motion compensation are considered. Later work on the subject which also included the option to change the DCT quantizer value on a macroblock to macroblock basis appeared in [SK96]. CHEN and WILLSON exploit dependencies in differential coding of motion vectors for motion estimation [CW98]. An example for the exploitation of temporal dependencies in video coding can be found in [ROV94]. The work of RAMCHANDRAN, ORTEGA, and VETTERLI in [ROV94] was extended by LEE and DICKINSON in [LD94], but video encoders for interactive communications must neglect this aspect to a large extent, since they cannot tolerate the delay necessary for optimizing a long temporal sequence of decisions.

In many coder control algorithms, including the one employed in this book, the spatial and temporal dependencies between blocks are neglected. This is because of the large parameter space involved and delay constraints. Hence, for each block $\boldsymbol{S}_k$, the coding mode with associated parameters is optimized given the decisions made for prior coded blocks. Consequently, the coding mode for each block is determined using the Lagrangian cost function in (2.3). This can easily be done for the INTRA coding mode via DCT transform and successive quantization as well as run-length encoding of the coefficients.

For the INTER coding mode, the associated parameter space is still very large and further simplifications are necessary. Ideally, decisions should be controlled by their ultimate effect on the resulting pictures, but this ideal may not be attainable or may not justify the associated complexity in all cases. Considering each possible motion vector to send for a picture area, an encoder should perform an optimized coding of the residual error and measure the resulting bit usage and distortion. Only by doing this can the best possible motion vector value be determined. However, there are typically thousands of possible motion vector values to choose from, and coding just one residual difference signal typically requires a significant fraction of the total computational power of a practical encoder.

A simple and widely accepted method of determining the Lagrangian costs for the INTER coding mode is to search for a motion vector that minimizes a Lagrangian cost criterion prior to residual coding. The bit-rate and distortion of the following residual coding stage are either ignored or approximated. Then, given the motion vector(s), the parameters for the residual coding stage are encoded. The minimization of a Lagrangian cost function for motion estimation as given in (2.3) was first proposed by SULLIVAN and BAKER [SB91]. A substantial amount of work on the subject has appeared in literature [CKS96, KLSW97, CW96, SK97, CW98].

A theoretical frame work for bit-allocation in a hybrid video coder has been introduced by GIROD [Gir94]. The analysis in [Gir94] provides the insight that the hybrid video coder should operate at constant distortion-rate slopes when allocating bits to the motion vectors and the residual coding.

2.3 CODER CONTROL FOR ITU-T RECOMMENDATION H.263

The ITU-T Video Coding Experts Group maintains a document describing examples of encoding strategies, called its test model. An important contribution of this book is the proposal of a coder control that is based on Lagrangian optimization techniques [ITU98b]. The proposal in [ITU98b] lead to the creation of a new test model: TMN-10 [ITU98d]. The TMN-10 coder control is used as a basis for comparison in this book to evaluate the proposed MCP ideas.

TMN-10 rate control utilizes macroblock mode decision similar to [WLM$^+$96] but without consideration of the dependencies of distortion and rate values on coding mode decisions made for past or future macroblocks. Hence, for each macroblock, the coding mode with associated parameters is optimized given the decisions made for prior coded blocks only. Consequently, the coding mode for each block is determined using the Lagrangian cost function in (2.3). Let the Lagrange parameter λ_{MODE} and the DCT quantizer value Q be given. The Lagrangian mode decision for a macroblock $\boldsymbol{S}_k$ in TMN-10 proceeds by minimizing

$$J_{\text{MODE}}(\boldsymbol{S}_k, I_k|Q, \lambda_{\text{MODE}}) = D_{\text{REC}}(\boldsymbol{S}_k, I_k|Q) + \lambda_{\text{MODE}} \cdot R_{\text{REC}}(\boldsymbol{S}_k, I_k|Q), \quad (2.7)$$

where the macroblock mode I_k is varied over the set $\{\text{INTRA}, \text{SKIP}, \text{INTER}, \text{INTER+4V}\}$. Rate $R_{\text{REC}}(\boldsymbol{S}_k, I_k|Q)$ and distortion $D_{\text{REC}}(\boldsymbol{S}_k, I_k|Q)$ for the various modes are computed as follows.

For the INTRA mode, the 8×8 blocks of the macroblock $\boldsymbol{S}_k$ are processed by a DCT and subsequent quantization. The distortion $D_{\text{REC}}(\boldsymbol{S}_k, \text{INTRA}|Q)$ is measured as the SSD between the reconstructed and the original macroblock pixels. The rate $R_{\text{REC}}(\boldsymbol{S}_k, \text{INTRA}|Q)$ is the rate that results after run-level variable-length coding.

For the SKIP mode, distortion $D_{\text{REC}}(\boldsymbol{S}_k, \text{SKIP})$ and rate $R_{\text{REC}}(\boldsymbol{S}_k, \text{SKIP})$ do not depend on the DCT quantizer value Q of the current picture. The distortion is determined by the SSD between the current picture and the previous coded picture for the macroblock pixels, and the rate is given as one bit per macroblock, as specified by ITU-T Recommendation H.263 [ITU96a].

The computation of the Lagrangian costs for the INTER and INTER+4V coding modes is much more demanding than for INTRA and SKIP. This is because of the block motion estimation step. The size of the blocks can be either 16×16 pixels for the INTER mode or 8×8 pixels for the INTER+4V mode. Let the Lagrange parameter λ_{MOTION} and the decoded reference picture $\acute{s}$ be given. Rate-constrained motion estimation for a block $\boldsymbol{S}_i$ is conducted by minimizing the Lagrangian cost function

$$\boldsymbol{m}_i = \underset{\boldsymbol{m} \in \boldsymbol{\mathcal{M}}}{\operatorname{argmin}} \left\{ D_{\text{DFD}}(\boldsymbol{S}_i, \boldsymbol{m}) + \lambda_{\text{MOTION}} R_{\text{MOTION}}(\boldsymbol{S}_i, \boldsymbol{m}) \right\}, \tag{2.8}$$

with the distortion term being given as

$$D_{\text{DFD}}(\boldsymbol{S}_i, \boldsymbol{m}) = \sum_{(x,y) \in \boldsymbol{\mathcal{A}}_i} |s[x, y, t] - \acute{s}[x - m_x, y - m_y, t - m_t]|^p \tag{2.9}$$

wit $p = 1$ for the sum of absolute differences (SAD) and $p = 2$ for the sum of squared differences (SSD). $R_{\text{MOTION}}(\boldsymbol{S}_i, \boldsymbol{m})$ is the bit-rate required for the motion vector. The search range $\boldsymbol{\mathcal{M}}$ is ± 16 integer pixel positions horizontally and vertically and the prior decoded picture is referenced ($m_t = 1$). Depending on the use of SSD or SAD, the Lagrange parameter λ_{MOTION} has to be adjusted as discussed in the next section. The motion search that minimizes (2.8) proceeds first over integer-pixel locations. Then, the best of those integer-pixel motion vectors is tested whether one of the surrounding half-pixel positions provides a cost reduction in (2.8). This step is regarded as half-pixel refinement and yields the resulting motion vector $\boldsymbol{m}_i$. The resulting prediction error signal $u[x, y, t, \boldsymbol{m}_i]$ is similar to the INTRA mode processed by a DCT and subsequent quantization. The distortion D_{REC} is also measured as the SSD between the reconstructed and the original macroblock pixels. The rate R_{REC} is given as the sum of the bits for the motion vector and the bits for the quantized and run-level variable-length encoded DCT coefficients.

The described algorithm is used within the ITU-T Video Coding Experts Group for the evaluation of rate-distortion performance. The TMN-10 specification also recommends utilizing the H.263+ Annexes D, F, I, J, and T [ITU98d]. To obtain rate-distortion curves, the coder is run with varying settings for the encoding parameters λ_{MODE}, λ_{MOTION}, and Q. A comparison that is based on this methodology has already been presented in Section 1.4 and is also employed in the following.

2.4 CHOOSING THE CODER CONTROL PARAMETERS

In this section, the selection of the coder control parameters λ_{MODE}, λ_{MOTION}, and Q is discussed. First, the experiment that leads to the proposed connection between these parameters is explained. Second, the relationship obtained for the

Lagrange parameters and DCT quantizer is interpreted. Finally, the efficiency of the proposed scheme is verified.

2.4.1 EXPERIMENTAL DETERMINATION OF THE CODER CONTROL PARAMETERS

In TMN-10, the Lagrange parameter λ_{MODE} controls the macroblock mode decision when evaluating (2.7). The Lagrangian cost function in (2.7) depends on the MCP signal and the DFD coding. The MCP signal is obtained by minimizing (2.8), which depends on the choice of λ_{MOTION}, while the DFD coding is controlled by the DCT quantizer value Q. Hence, for a fixed value of λ_{MODE}, a certain setting of λ_{MOTION} and Q provides the optimal results in terms of coding efficiency within the TMN-10 framework. One approach to find the optimal values of λ_{MOTION} and Q is to evaluate the product space of these two parameters. For that, each pair of λ_{MOTION} and Q has to be considered that could provide a minimum Lagrangian cost function in (2.7). However, this approach requires a prohibitive amount of computation. Therefore, the relationship between λ_{MODE} and Q is considered first while fixing λ_{MOTION}. The parameter λ_{MOTION} is adjusted according to $\lambda_{\text{MOTION}} = \lambda_{\text{MODE}}$ when considering the SSD distortion measure in (2.8). This choice is motivated by theoretical [Gir94] and experimental results that are presented later in this section.

To obtain a relationship between Q and λ_{MODE}, the minimization of the Lagrangian cost function in (2.7) is extended by the macroblock mode type INTER+Q, which permits changing Q by a small amount when sending an INTER macroblock. More precisely, the macroblock mode decision is conducted by minimizing (2.7) over the set of macroblock modes

$$\{\text{INTRA}, \text{SKIP}, \text{INTER}, \text{INTER+4V}, \ldots \tag{2.10}$$
$$\text{INTER+Q}(-2), \text{INTER+Q}(-1), \text{INTER+Q}(+1), \text{INTER+Q}(+2)\},$$

where, for example, INTER+Q(-2) stands for the INTER macroblock mode being coded with DCT quantizer value reduced by two relative to the previous macroblock. Hence, the Q value selected by the minimization routine becomes dependent on λ_{MODE}. Otherwise the algorithm for running the rate-distortion optimized video coder remains unchanged from the TMN-10 specification in Section 2.3.

Figure 2.1 shows the relative frequency of chosen macroblock quantizer values Q for several values of λ_{MODE}. The Lagrange parameter λ_{MODE} is varied over seven values: 4, 25, 100, 250, 400, 730, and 1000, producing seven normalized histograms for the chosen DCT quantizer value Q that are depicted in the plots in Fig. 2.1. In Fig. 2.1, the macroblock Q values are gathered while coding 100 frames of the video sequences *Foreman*, *Mobile & Calendar*, *Mother & Daughter*, and *News*. The quantizer value Q does not vary much given a fixed

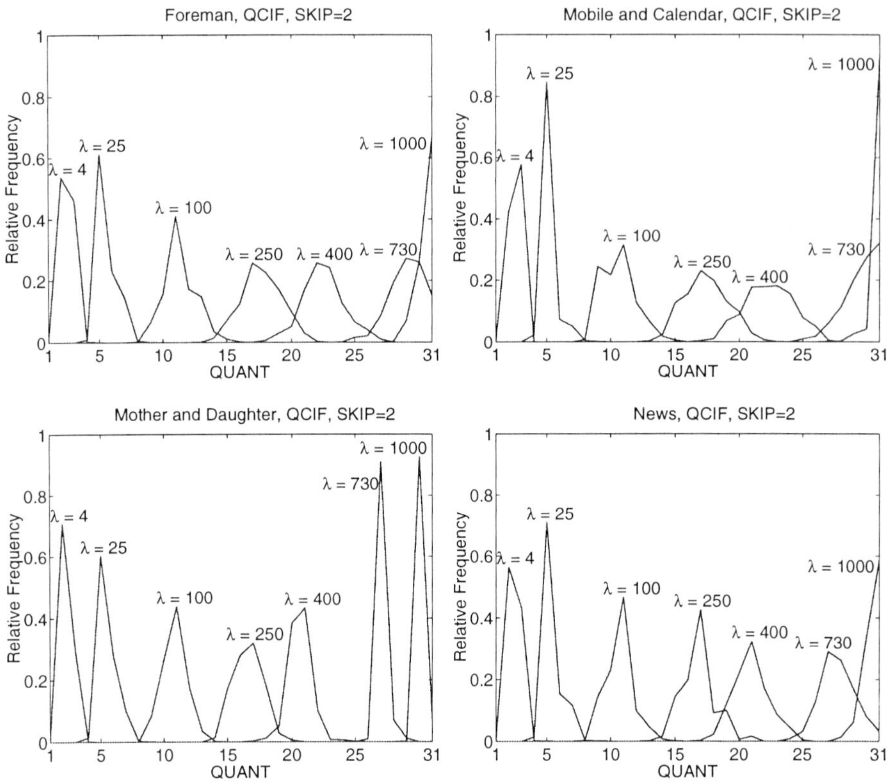

Figure 2.1. Relative frequency vs. macroblock Q for various values of the Lagrange parameter λ_{MODE}. The relative frequencies of macroblock Q values are gathered while coding 100 frames of the video sequences *Foreman* (top left), *Mobile & Calendar* (top right), *Mother & Daughter* (bottom left), and *News* (bottom right).

value of λ_{MODE}. Moreover, as experimental results show, the gain when permitting the variation is rather small, indicating that fixing Q as in TMN-10 might be justified.

As can already be seen from the histograms in Fig. 2.1, the peaks of the histograms are very similar among the four sequences and they are only dependent on the choice of λ_{MODE}. This observation can be confirmed by looking at the left-hand side of Fig. 2.2, where the average macroblock quantizer values Q from the histograms in Fig. 2.1 are shown. The bold curve in Fig. 2.2 depicts the function

$$\lambda_{\text{MODE}}(Q) \approx 0.85 \cdot Q^2 \,, \tag{2.11}$$

which is an approximation of the relationship between the macroblock quantizer value Q and the Lagrange parameter λ_{MODE} up to Q values of 25. H.263 allows only a choice of $Q \in \{1, 2, \ldots, 31\}$. In the next section, a motivation is given for the relationship between Q and λ_{MODE} in (2.11).

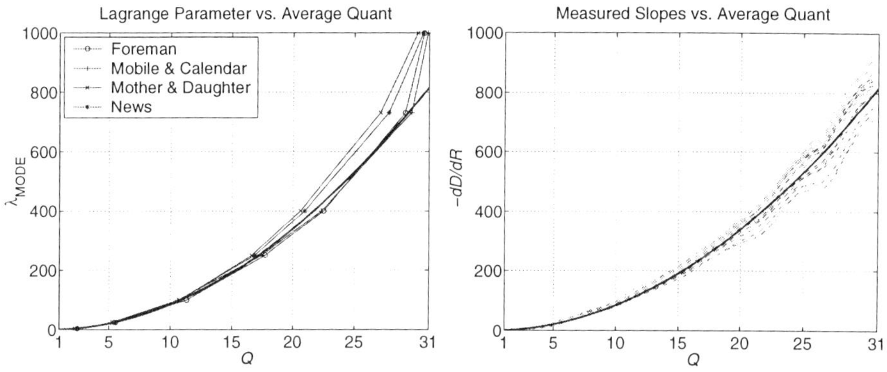

Figure 2.2. Lagrange parameter λ_{MODE} vs. average macroblock Q (left) and measured slopes (right).

2.4.2 INTERPRETATION OF THE LAGRANGE PARAMETER

The Lagrange parameter is regarded as the negative slope of the distortion-rate curve [Eve63, SG88, CLG89]. It is simple to show that if the distortion-rate function $D_{\mathrm{REC}}(R_{\mathrm{REC}})$ is strictly convex then $J_{\mathrm{MODE}}(R_{\mathrm{REC}}) = D_{\mathrm{REC}}(R_{\mathrm{REC}}) + \lambda_{\mathrm{MODE}} R_{\mathrm{REC}}$ is strictly convex as well. Assuming $D_{\mathrm{REC}}(R_{\mathrm{REC}})$ to be differentiable everywhere, the minimum of the Lagrangian cost function is given by setting its derivative to zero, i.e.

$$\frac{\mathrm{d}J_{\mathrm{MODE}}}{\mathrm{d}R_{\mathrm{REC}}} = \frac{\mathrm{d}D_{\mathrm{REC}}}{\mathrm{d}R_{\mathrm{REC}}} + \lambda_{\mathrm{MODE}} \stackrel{!}{=} 0, \tag{2.12}$$

which yields

$$\lambda_{\mathrm{MODE}} = -\frac{\mathrm{d}D_{\mathrm{REC}}}{\mathrm{d}R_{\mathrm{REC}}}. \tag{2.13}$$

A typical high-rate approximation curve for entropy-constrained scalar quantization can be written as [JN94]

$$R_{\mathrm{REC}}(D_{\mathrm{REC}}) = a \log_2 \left(\frac{b}{D_{\mathrm{REC}}} \right), \tag{2.14}$$

with a and b parameterizing the functional relationship between rate and distortion. For the distortion-to-quantizer relation, it is assumed that at sufficiently high rates, the source probability distribution can be approximated as uniform within each quantization interval [GP68] yielding

$$D_{\mathrm{REC}} = \frac{(2Q)^2}{12} = \frac{Q^2}{3}. \tag{2.15}$$

Note that in H.263 the macroblock quantizer value Q is approximately double the distance of the quantizer reproduction levels. The total differentials of rate and distortion are given as

$$\mathrm{d}R_{\mathrm{REC}} = \frac{\partial R_{\mathrm{REC}}}{\partial Q}\mathrm{d}Q = \frac{-2a}{Q\ln 2}\mathrm{d}Q \qquad \text{and} \qquad \mathrm{d}D_{\mathrm{REC}} = \frac{\partial D_{\mathrm{REC}}}{\partial Q}\mathrm{d}Q = \frac{2Q}{3}\mathrm{d}Q \tag{2.16}$$

Plugging these into (2.13), provides the result

$$\lambda_{\mathrm{MODE}}(Q) = -\frac{\mathrm{d}D_{\mathrm{REC}}(Q)}{\mathrm{d}R_{\mathrm{REC}}(Q)} = c \cdot Q^2 \tag{2.17}$$

where $c = \ln 2/(3a)$. Although the assumptions here may not be completely realistic, the derivation reveals at least the qualitative insight that it may be reasonable for the value of the Lagrange parameter λ_{MODE} to be proportional to the square of the quantizer value. As shown above by means of experimental results, 0.85 appears to be a reasonable value for use as the constant c.

For confirmation of the relationship in (2.17), an experiment has been conducted to measure the distortion-rate slopes $\mathrm{d}D_{\mathrm{REC}}(Q)/\mathrm{d}R_{\mathrm{REC}}(Q)$ for a given value of Q. The experiment consists of the following steps:

1. The TMN-10 coder is run employing quantizer values $Q_{\mathrm{REF}} \in \{4, 5, 7, 10, 15, 25\}$. The resulting bit-streams are decoded and the reconstructed frames are employed as reference frames in the next step.

2. Given the coded reference frames, the MCP signal is computed for a fixed value of

$$\lambda_{\mathrm{MOTION}} = 0.85 \cdot Q_{\mathrm{REF}}^2 \tag{2.18}$$

   when employing the SSD distortion measure in the minimization of (2.8). Here, only 16×16 blocks are utilized for half-pixel accurate motion compensation. The MCP signal is subtracted from the original signal providing the DFD signal that is further processed in the next step.

3. The DFD signal is encoded for each frame when varying the value of the DCT quantizer in the range $Q = \{1, \ldots, 31\}$ for the INTER macroblock mode. The other macroblock modes have been excluded here to avoid the macroblock mode decision that involves Lagrangian optimization using λ_{MODE}.

4. For each sequence and Q_{REF}, the distortion and rate values per frame including the motion vector bit-rate are averaged, and the slopes are computed numerically.

Via this procedure, the relationship between the DCT quantizer value Q and the slope of the distortion-rate curve $\mathrm{d}D_{\mathrm{REC}}(Q)/\mathrm{d}R_{\mathrm{REC}}(Q)$ has been obtained as shown on the right-hand side of Fig. 2.2. This experiment shows that the relationship in (2.17) can be measured using the rate-distortion curve for the DFD coding part of the hybrid video coder. This is in agreement with the experiment that is employed to establish (2.11).

For further interpretation, an experiment is conducted, which yields the distortion-rate slopes as well as the functional relationship between λ_{MODE} and Q when permitting all macroblock modes, i.e., SKIP, INTRA, INTER, INTER+4V. For that, the above algorithm is repeated where steps 2 and 3 have been modified to

2. Given the coded reference frames, 2 MCP signals are computed for a fixed value of
$$\lambda_{\mathrm{MOTION}} = 0.85 \cdot Q_{\mathrm{REF}}^2 \tag{2.19}$$
when employing the SSD distortion measure in the minimization of (2.8). For the INTER macroblock mode, 16×16 blocks are utilized for half-pixel accurate motion compensation, while for the INTER+4V macroblock mode, 8×8 blocks are employed. For the SKIP macroblock mode, the coded reference frame is used as the prediction signal. The prediction signals for the three macroblock modes are subtracted from the original signal providing the DFD signals that are further processed in the next step.

3. Lagrangian costs are computed for the macroblock modes $\{\mathrm{INTRA}, \mathrm{SKIP}, \mathrm{INTER}, \mathrm{INTER{+}4V}\}$ and for each value of $Q \in \{1, \ldots, 31\}$ given the MCP signal. Given the costs for all cases, the macroblock mode decision is conducted by minimizing (2.7), where λ_{MODE} is adjusted by (2.11) using Q.

Figure 2.3 shows the result for the described experiment for the sequences *Foreman*, *Mobile & Calendar*, *Mother & Daughter*, and *News*. In Fig. 2.3, the relationship between the slope and the DCT quantizer in (2.17) is used to obtain a prediction for the distortion given a measurement of the bit-rate. Thus, the solid curves in Fig. 2.3 correspond to this distortion prediction. Each curve corresponds to one value of the quantizer for the reference picture $Q_{\mathrm{REF}} \in \{4, 5, 7, 10, 15, 25\}$. The distortion prediction is conducted via approximating

$$\frac{dD_{\mathrm{REC}}(Q+0.5)}{dR_{\mathrm{REC}}(Q+0.5)} \approx \frac{D_{\mathrm{REC}}(Q+1) - D_{\mathrm{REC}}(Q)}{R_{\mathrm{REC}}(Q+1) - R_{\mathrm{REC}}(Q)} \approx -0.85 \cdot (Q+0.5)^2. \tag{2.20}$$

A simple manipulation yields an iterative procedure for the prediction of the distortion

$$D_{\mathrm{REC}}(Q+1) = D_{\mathrm{REC}}(Q) + 0.85 \cdot (Q+0.5)^2 \cdot [R_{\mathrm{REC}}(Q) - R_{\mathrm{REC}}(Q+1)]. \tag{2.21}$$

The points marked with a star correspond to measured distortion $D_{\mathrm{REC}}(Q)$ and bit-rate $R_{\mathrm{REC}}(Q)$ for DCT quantizer value $Q = 4$. These points are used to

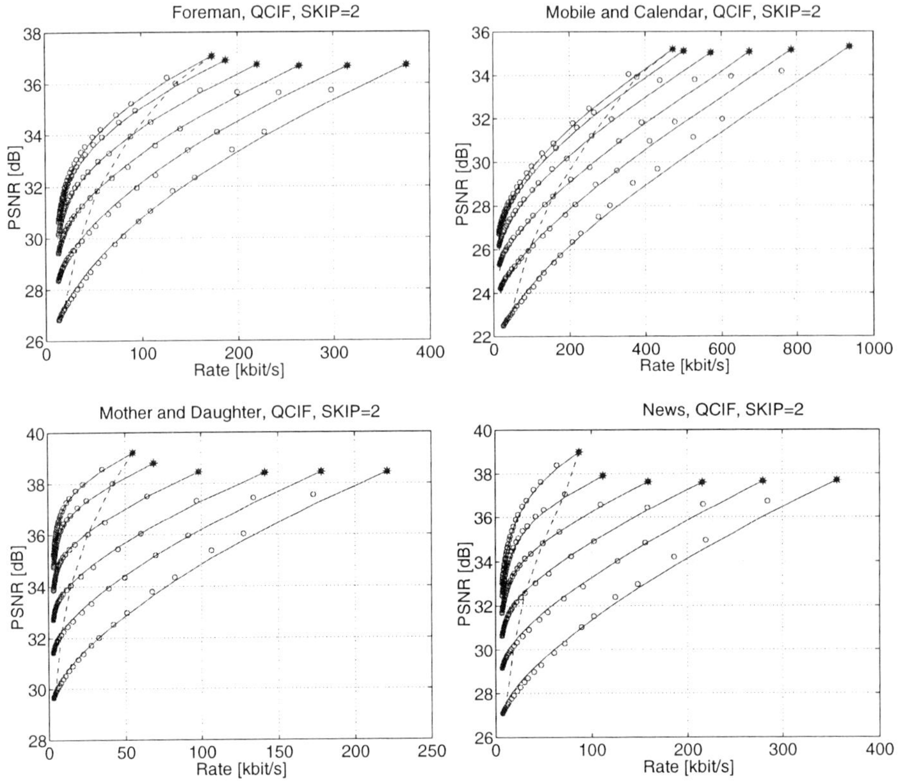

Figure 2.3. PSNR in dB vs. bit-rate in kbit/s for the video sequences *Foreman* (top left), *Mobile & Calendar* (top right), *Mother & Daughter* (bottom left), and *News* (bottom right).

initialize the iterative procedure to predict distortion via (2.21) given the measured bit-rates for all values of Q. For all measurements in Fig. 2.3, distortion corresponds to average mean squared error (MSE) that is first measured over the sequence and then converted into PSNR versus average overall bit-rate in kbits/s. The circles correspond to distortion and bit-rate measurements for the cases $D_{\mathrm{REC}}(Q)$ and bit-rate $R_{\mathrm{REC}}(Q)$ with $Q \in \{5, \ldots, 31\}$. The measured and predicted distortion values are well aligned validating that the slope-quantizer relationship in (2.17) is correct and that theses slopes can indeed be measured for the solid curves. As a comparison, the dashed lines connect the rate-distortion curves for the case that the DCT quantizer of the reference picture Q_{REF} is the same as the DCT quantizer of the coded macroblocks Q.

The functional relationship in (2.17) as depicted in Fig. 2.2 also describes the results from similar experiments when varying temporal or spatial resolution and give further confirmation that the relationship in (2.17) provides an acceptable characterization of the DCT-based DFD coding part of the hybrid video coder.

2.4.3 EFFICIENCY EVALUATION FOR THE PARAMETER CHOICE

The choice of the encoding parameters has to be evaluated based on its effect on rate-distortion performance. Hence, in order to verify that the particular choice of the relationship between λ_{MODE}, λ_{MOTION}, and Q provides good results in rate-distortion performance, the H.263+ coder is run using the TMN-10 algorithm for the product space of the parameter sets $\lambda_{\text{MODE}}, \lambda_{\text{MOTION}} \in \{0, 4, 14, 21, 42, 85, 191, 531, 1360, 8500\}$ and $Q \in \{4, 5, 7, 10, 15, 25\}$. For each of the 600 combinations of the three parameters, the sequences *Foreman*, *Mobile & Calendar*, *Mother & Daughter*, and *News* are encoded, and the resulting average rate-distortion points are depicted in Fig. 2.4. The rate-distortion

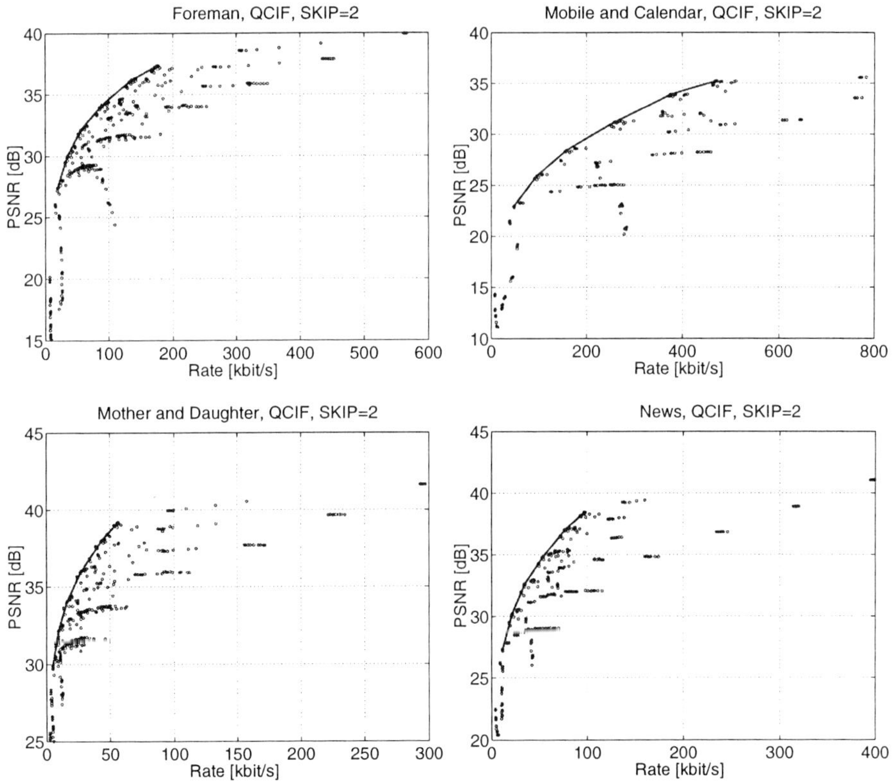

Figure 2.4. PSNR in dB vs. bit-rate in kbit/s when running TMN-10 with various λ_{MODE}, λ_{MOTION}, and Q combinations for the video sequences *Foreman* (top left), *Mobile & Calendar* (top right), *Mother & Daughter* (bottom left), and *News* (bottom right).

points obtained when setting $\lambda_{\text{MODE}} = \lambda_{\text{MOTION}} = 0.85Q^2$ are connected by the line in Fig. 2.4 and indicate that this setting indeed provides good results for all tested sequences. Although not shown here, it has been found that also for other sequences as well as other temporal and spatial resolutions, similar results can be obtained.

So far, SSD has been used as distortion measure for motion estimation. In case SAD is used for motion estimation, λ_{MOTION} is adjusted as

$$\lambda_{\text{MOTION}} = \sqrt{\lambda_{\text{MODE}}}. \tag{2.22}$$

Using this adjustment, experiments show that both distortion measures SSD and SAD provide very similar results.

2.5 COMPARISON TO OTHER ENCODING STRATEGIES

TMN-9 [ITU97] is the predecessor to TMN-10 as the recommended encoding algorithm for H.263+. The TMN-9 mode decision method is based on thresholds. A cost measure for the INTRA macroblock mode containing pixels in the set $\mathcal{A}_k$ is computed as

$$C_{\text{INTRA}} = \sum_{(x,y)\in\mathcal{A}_k} |s[x,y,t] - \mu_{\mathcal{A}_k}| \tag{2.23}$$

with $\mu_{\mathcal{A}_k}$ being the mean of the pixels of the 16×16 macroblock. For the INTER macroblock mode, the cost measure is given as

$$C_{\text{INTER}}(\mathcal{M}^F) = \tag{2.24}$$
$$\min_{(m_x,m_y)\in\mathcal{M}^F} \sum_{(x,y)\in\mathcal{A}_k} |s[x,y,t] - \acute{s}[x-m_x, y-m_y, t-m_t]| - \xi(m_x,m_y)$$

where the motion search proceeds only over the set of integer-pixel (or full-pixel) positions $\mathcal{M}^F = \{-16\ldots16\} \times \{-16\ldots16\}$ in the previous decoded frame yielding the minimum SAD value and the corresponding motion vector $\boldsymbol{m}_k^F$. If the x and y component of the motion vector $\boldsymbol{m}$ are zero, the value of ξ is set to 100 to give a preference towards choosing the SKIP mode. Otherwise, ξ is set to 0. Given the two cost measures in (2.23) and (2.24), the following inequality is evaluated

$$C_{\text{INTRA}} < C_{\text{INTER}}(\mathcal{M}^F) - 500 \tag{2.25}$$

When this inequality is satisfied, the INTRA mode is chosen for the macroblock and transmitted.

If the INTER mode is chosen, the integer-pixel motion vector $\boldsymbol{m}_k^F$ is used as the initialization of the half-pixel motion estimation step. For that the cost measure in (2.24) is employed but the set of integer-pixel locations $\mathcal{M}^F$ is replaced by the set of half-pixel locations $\mathcal{M}^H(\boldsymbol{m}_k^F)$ that surround $\boldsymbol{m}_k^F$. This step yields the cost measure $C_{\text{INTER}}(\mathcal{M}^H(\boldsymbol{m}_k^F))$. The four motion vectors for the 8×8 blocks of the INTER+4V mode are found as well by utilizing (2.24) when replacing $\mathcal{M}^F$ with $\mathcal{M}^H(m_k^F)$. But here, the set of pixels for the SAD computation is changed to the 8×8 blocks yielding the cost measure $C_{\text{INTER+4V},l}(\mathcal{M}^H(m_k^F))$ for the lth block. The INTER+4V mode is chosen if

$$\sum_{l=1}^{4} C_{\text{INTER+4V},l}(\mathcal{M}^H(m_k^F)) < C_{\text{INTER}}(\mathcal{M}^H(m_k^F)) - 200. \tag{2.26}$$

is satisfied. The SKIP mode is chosen in TMN-9 only if the INTER mode is preferred to the INTRA mode and the motion vector components and all of the quantized transform coefficients are zero.

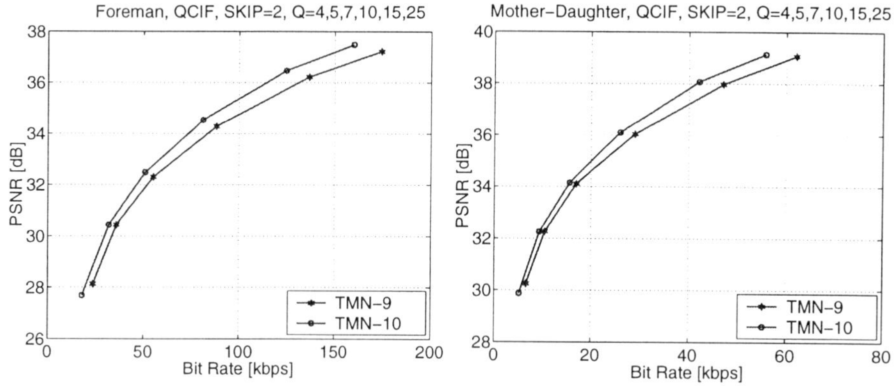

Figure 2.5. Coding performance for the sequence *Foreman* (left) and *Mother & Daughter* (right) when comparing the encoding strategies of TMN-9 and TMN-10.

The role of the encoding strategy is demonstrated in Fig. 2.5 for the video sequences *Foreman* and *Mother & Daughter*. For both curves, the same bit-stream syntax is used, with changes only in the mode decision and motion estimation strategies, either TMN-9 or TMN-10. The overall performance gain of TMN-10 is typically between 5 and 10% in bit-rate when comparing at a fixed reconstruction quality of 34 dB PSNR.

2.6 CHAPTER SUMMARY

Given a set of source samples and a rate constraint, Lagrangian optimization is a powerful tool for bit-allocation that can be applied to obtain a set of either dependent or independent coding options. When the coding options depend on each other, the search has to proceed over the product space of coding options and associated parameters which in most cases requires a prohibitive amount of computation. Some dependencies are trellis-structured and researchers have indeed used dynamic programming methods in combination with Lagrangian bit-allocation to exploit those dependencies between DCT coefficients or between blocks. But the optimization task still remains rather unwieldy because of the large amount of computation involved. Hence, in most practical systems, the dependencies between blocks are ignored and decisions are made assuming the encoding of past parameters as being fixed.

A practical and widely accepted optimization approach to hybrid video coding is to use rate-constrained motion estimation and mode decision that are conducted for each block independently. TMN-10, the encoder recommendation for H.263+ specifies such a strategy. TMN-10 has been created by the ITU-T Video Coding Experts Group based on a contribution of this book. The contribution is an efficient approach for choosing the encoding parameters which has been for a long time an obstacle for the consideration of Lagrangian coder control in practical systems. The comparison to TMN-9 shows that a bit-rate reduction up to 10 % can be achieved. The strategy in TMN-9 is based on heuristics and thresholds. The performance and generality of the TMN-10 coder control make the approach suitable for controlling more sophisticated video coders as well, as proposed in the next chapters.

Chapter 3

LONG-TERM MEMORY MOTION-COMPENSATED PREDICTION

In most existing video codecs, motion compensation is carried out by referencing the prior decoded frame. So far, multiple reference frames have been considered only to a very limited extent for two reasons. *First*, they simply could not be afforded. However, the continuously dropping costs of semiconductors are making the storage and processing of multiple reference frames possible. *Second*, it was not believed that multiple reference frames would significantly improve coding efficiency. In this chapter, methods for multi-frame MCP are investigated and it is shown that significant improvements in coding efficiency are, in fact, possible.

Multi-frame MCP extends the motion vector utilized in block-based motion compensation by a picture reference parameter to employ more frames than the prior decoded one. The purpose of that is to improve rate-distortion performance. The picture reference parameter is transmitted as side information requiring additional bit-rate. An important question is which reference pictures are efficient in terms of rate-distortion performance. In general, any useful image data may be utilized as reference frames. An important rule is that the bit-rate overhead that is due to employing a particular reference frame must be lower than the bit-rate savings. For that, rate-constrained motion estimation and mode decision are utilized to control the bit-rate.

One simple and efficient approach is to utilize past decoded frames as reference pictures since they are available at encoder and decoder simultaneously at practically no bit-rate overhead. The idea behind this approach is to exploit long-term statistical dependencies and therefore, the name *long-term memory MCP* has been coined for it in [WZG99] where parts of this chapter have been published before. Encoder and decoder negotiate versus a memory control that the multi-frame buffer covers several decoded frames simultaneously. In addition to a spatial displacement, the motion estimation also determines for each block which picture to reference. Hence, for long-term memory MCP the picture reference parameter relates to a time delay.

In this chapter, block-based hybrid video compression using long-term memory MCP is investigated and the practical approaches together with the results that lead to the incorporation of long-term memory MCP into the ITU-T Recommendation H.263++ via Annex U are described [ITU00]. In Section 3.1, the block diagram of the long-term memory motion-compensating predictor is presented and the various buffering modes that can also be found in Annex U of H.263++ are explained. The effects that cause the improved prediction performance of long-term memory MCP are analyzed in Section 3.2. A statistical model that describes the prediction gains is given in Section 3.3. Section 3.4 describes how long-term memory prediction can be integrated into a hybrid video codec. The performance of an H.263+ coder compared against an H.263++ coder incorporating long-term memory MCP via Annex U is compared by means of experimental results.

3.1 LONG-TERM MEMORY MOTION COMPENSATION

The block diagram of a long-term memory motion-compensated predictor is shown in Fig. 3.1. It shows a motion-compensated predictor that can utilize M frame memories, with $M \geq 1$. The *memory control* is used to arrange the reference frames. The MCP signal $\hat{s}$ is generated via block-based *multi-frame motion compensation* where for each block one of several previous decoded frames $\acute{s}$ is indicated as a reference. For that, the spatial displacement vector (m_x, m_y) is extended by a picture reference parameter m_t which requires additional bit-rate in case $M > 1$. The motion vectors are determined by *multi-frame motion estimation* which is conducted via block matching on each frame memory.

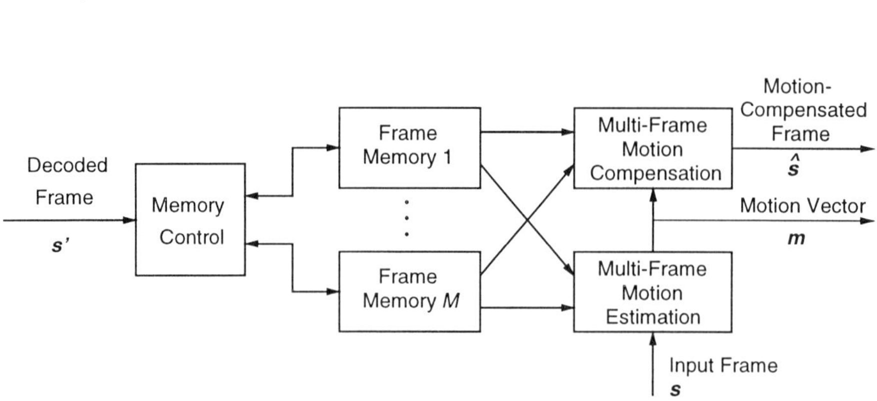

Figure 3.1. Multi-Frame Motion-Compensated Predictor.

The memory control arranges the reference frames according to a scheme that is shared by encoder and decoder. Such a scheme is important, because the picture reference parameter functions as a relative buffer index and a buffer mismatch would result in different MCP signals at encoder and decoder. Moreover, the memory control is designed to enable a custom arrangement of reference frames given a fixed variable length code for the transmission of the picture reference parameter. The variable length code assigns short code words to small values of the picture reference parameter and long code words to large values. The picture reference parameter is transmitted for each motion vector, such that the arrangement by the memory control has a significant impact on the behavior of the video codec regarding rate-distortion performance, computational complexity and memory requirements.

In general, several modes of operation for the memory control may be defined and the one which is used may be negotiated between encoder and decoder. In this work, the following schemes are proposed for memory control:

1. *Sliding Window:* The reference pictures are arranged and indexed on a first-in-first-out basis. For that, past decoded and reconstructed frames starting with the prior decoded frame ending with the frame that is decoded M time instants before are collected in the frame memories 1 to M.

2. *Index Mapping:* Modification of the indexing scheme for the multi-frame buffer. The physical structure of the multi-frame buffer is unchanged. Only the meaning of the picture reference parameter for each motion vector is modified according to the *Index Mapping* scheme.

3. *Adaptive Buffering:* The arrangement of the reference frames can be varied on a frame-by-frame basis. For each decoded frame, an indication is transmitted whether this picture is to be included in the multi-frame buffer. Moreover, another indication is used to specify which picture has to be removed from the multi-frame buffer.

The first approach to memory control, *Sliding Window*, is straightforward conceptually, since the most recent decoded frames in many natural camera-view scenes are also very likely to contain useful prediction material. If a fixed number of frames M is used, the *Sliding Window* approach minimizes the time at the beginning of the sequence to exploit the full memory size since it accepts each decoded frame as reference. Also the variable length code used to index the reference frames follows the statistics of frame selections for natural camera-view scenes. Figure 3.2 illustrates the motion compensation process for the *Sliding Window* approach for the case of $M = 3$ reference frames. For each block, one out of M frames, which are the most recently decoded frames, can be referenced for motion compensation. Many results with long-term memory prediction in this chapter are obtained employing the *Sliding Window* memory control.

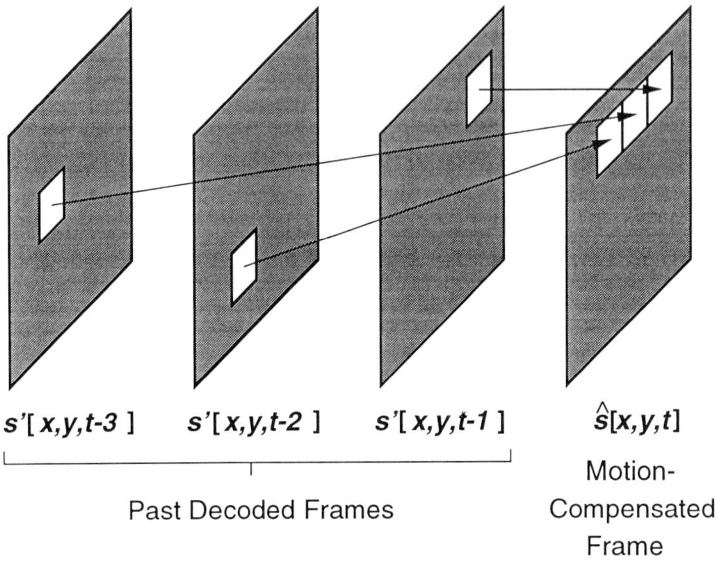

Figure 3.2. Long-term memory motion compensation.

As an alternative, the set of past decoded and reconstructed frames may be temporally sub-sampled. For that, the memory control options 2 and 3 are proposed which have different advantages and they are indeed used in various applications as will be described later. The *Index Mapping* scheme leaves the set of reference frames physically intact but changes their addressing. This scheme provides an option to adapt the ordering of the frame indices to the selection statistics on a frame-by-frame basis and therefore can provide bit-rate savings for the picture reference parameter that is transmitted with each motion vector. An application of this scheme is given in Chapter 4, where long-term memory prediction is combined with affine motion models and in Chapter 6, where long-term memory prediction is used for robust video transmission over error-prone channels.

The last approach, *Adaptive Buffering*, changes the set of buffered frames in that a decoded frame may not be included into the multi-frame buffer or that a particular frame is removed from the buffer. This scheme maybe used to lower the memory requirements. An application of this approach is presented in this chapter, which is a surveillance sequence.

3.2 PREDICTION PERFORMANCE

Improvements when using long-term memory MCP can be expected in case of a repetition of image sequence content that is captured by the multi-frame buffer. Note that such a repetition may or may not be meaningful in terms of human visual perception. Examples for such an effect are:

1. scene cuts,
2. uncovered background,
3. texture with aliasing,
4. similar realizations of a noisy image sequence.

In the following, the prediction gains that can be obtained with long-term memory MCP for these effects are illustrated.

3.2.1 SCENE CUTS

Scene cuts can provide very substantial gains which are well exploited by long-term memory prediction. One example is a surveillance sequence that consists of several different sub-sequences, which are temporally interleaved [ITU98e]. Figure 3.3 shows reconstruction quality in PSNR vs. time in camera switch cycles. Each camera switch cycle corresponds to 4 seconds of video that

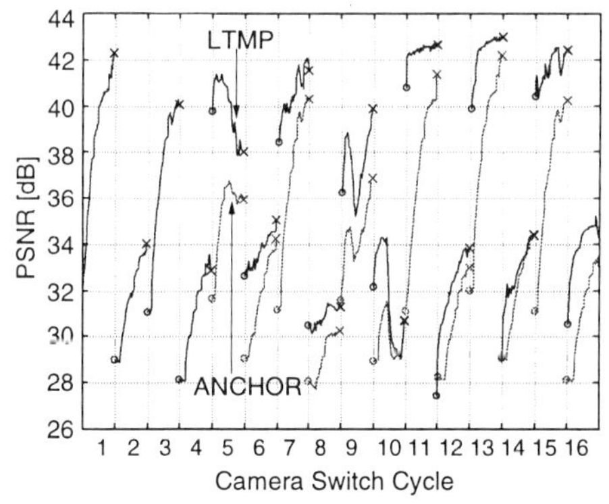

Figure 3.3. Results for surveillance sequence.

are captured by the same camera. Then, a switch occurs to the next camera depicting different content. In the application considered here, we utilize 4 cameras thus cycling through all of them takes 16 seconds. This cycling through the 4 cameras is repeated 3 times. Hence, the time axis in Fig. 3.3 shows 16 cycles which correspond to 64 seconds of video. This is a typical setting in such surveillance applications. Two codecs are compared which are

- **ANCHOR:** The H.263+ codec, which transmits an INTRA-frame when switching to a different camera.

- **LTMP:** The long-term memory codec when using the *Adaptive Buffering* memory control. In case a camera switch occurs, the last reconstructed frame from camera n is stored and can be referenced for prediction when the output of camera n is shown again. The long-term memory codec gets an external indication when such a camera switch occurs and it uses the *Adaptive Buffering* memory control to arrange the reference frames. In case no camera switch occurs, the same approach is taken as for the anchor, which is to reference the prior decoded frame.

Both codecs follow the TMN-10 coder control as described in Section 2.3. They utilize Annexes D, F, I, J, and T [ITU98a]. The bit-rate is controlled via varying the DCT quantizer step size Q so as to obtain 12.5 kbit/s. The coder control can skip frames in case the bit-rate computed for the maximum value of $Q = 31$ is exceeded. For the first 4 cycles, both codecs perform identically. Then, in the fifth cycle, the long-term memory MCP coder can use the last decoded frame from the first cycle providing a PSNR gain up to 2-8 dB and also higher temporal resolution. This benefit can be exploited at the beginning of all succeeding cycles as well.

Besides the surveillance application as described above, long-term memory prediction can also be beneficially employed for improved coding of other video source material with the same underlying structure. Another example is an interview scene where two cameras switch between the speakers. Other researchers have also shown that the *Adaptive Buffering* memory control syntax if combined with a scene change detector can provide substantial gains for the scene cuts [ZK99a, ITU99a].

The gain for the surveillance application is obtained via adapting the memory control to the structure of the underlying video capture on the frame basis. In the following, an example is given where the gain is obtained when using multiple reference frames to compensate each macroblock or block. Such an example provides the sequence *News*. This MPEG-4 test sequence is an artificially constructed sequence of 10 seconds in QCIF resolution. In the background of the sequence, two distinct sequences of dancers are displayed of which a still image of one sequence is shown in the left-hand side picture of Fig. 3.4. These sequences, however, are repeated every 5 seconds corresponding to 50 frames in the long-term memory buffer when sampling at 10 frames/s. Hence, in case the long-term memory coder can reference the frame that has been coded 50 time instances in the past on the macroblock or block basis, this effect can be beneficially exploited.

The right-hand side picture of Fig. 3.4 shows the average ΔPSNR gains per macroblock. The ΔPSNR gains are obtained for each macroblock as the

Figure 3.4. ΔPSNR gains for the MCP error per macroblock for *News*. The left-hand side plot shows the first frame of the sequence *News*, while the right-hand side plot shows ΔPSNR gains that are superimposed on the picture for each macroblock.

difference between the PSNR values for the MCP error when utilizing 50 reference frames and 1 reference frame. MCP is conducted via estimating motion vectors by minimizing SSD when considering 16×16 blocks that are ± 16 pixels spatially displaced in horizontal and vertical direction. For the case of long-term memory MCP, each of the 50 reference frames is searched while the memory control assembles the reference frames using the *Sliding Window* buffering mode. The results are obtained for frames $150, 153, ..., 297$ of the *News* sequence when using original frames as reference. The luminance value inside the picture on the right-hand side of Fig. 3.4 corresponds to the average squared frame difference of the pixels for those frames. The grid shows the macroblock boundaries and the numbers correspond to the difference in PSNR. The area that covers the dancer sequence shows very large PSNR gains up to 20 dB. These gains also extend to the case when referencing decoded pictures and considering the bit-rate of the picture reference parameter as shown later in this chapter. It is also worth noting that gains are obtained for other parts of the picture which do not result from the scene cut. Long-term memory prediction on the block-basis permits to exploit those gains as well.

3.2.2 UNCOVERED BACKGROUND

The prediction gain due to uncovered background effects is illustrated for the sequence *Container Ship* in Fig. 3.5. The lower part of the left-hand side picture visualizes two birds that fly through the scene. The picture is constructed via superimposing the frames $150 \ldots 299$ of the sequence to show the trajectory that the birds cover in the image sequence. This trajectory can be found again in the right-hand side plot of Fig. 3.5. This plot shows the PSNR gains per macroblock that were obtained by a similar method as for the results in Fig. 3.4.

But here, the sequence *Container Ship* is employed and the long-term memory case utilizes only 5 reference frames instead of 50. The PSNR gains follow the trajectory of the birds. Since the image portion that the two objects cover is rather small, those gains are also comparably small because of the averaging. Nevertheless, the uncovered background effect is very important since it occurs in many scenes.

Figure 3.5. Sequence and PSNR gains per macroblock for the sequence *Container Ship* (right).

3.2.3 TEXTURE WITH ALIASING

The prediction gains that can be obtained when texture with aliasing occurs are illustrated for the sequence *Container Ship* as well. Please refer to the upper part of the pictures in Fig. 3.5. There, a ship is shown that is moving from left to right during the sequence. The superstructure on the ship contains high resolution texture. The two macroblocks with a bold frame show quite significant gains in PSNR which are around 9 dB. These gains are obtained for long-term memory prediction by employing integer-pixel motion vectors that reference the frame which is 3 time instants in the past. Note that for these two macroblocks of the sequence *Container Ship*, the long-term memory predictor never estimates half-pixel motion vectors which involve bilinear interpolation. The single-frame prediction anchor utilizes half-pixel motion vectors to compensate those macroblocks. These facts suggest that the gain here is obtained by referencing high resolution texture with aliasing providing improved prediction results also in case the reference picture is coded at a good quality. It should also be noted that the high-frequency superstructure also requires highly accurate motion vectors and that such sub-pixel motion may also be represented by the long-term memory approach.

3.2.4 SIMILAR REALIZATIONS OF A NOISY IMAGE SEQUENCE

The last effect which is caused by the occurrence of two similar realizations of a noisy image sequence is illustrated in Fig. 3.6 by means of the sequence *Silent Voice*. The left-hand side picture shows the result of the above described prediction experiment where the 10 reference frames are original frames, while the right-hand side corresponds to the case that the reference pictures are quantized. The part of the image sequence that is always background is framed by bold lines in both pictures of Fig. 3.6. This part is static throughout the sequence and hence the gains between 0.5 to 1 dB in the background part of left-hand side picture are obtained by referencing a similar realization of the noisy image sequence in the long-term memory buffer. An indication that this interpretation is correct is shown in the right-hand side picture, where the quantization of the reference pictures almost completely removes those gains.

1.1	1.1	1.2	1.1	0.2	1.0	0.6	1.0	0.9	0.8	1.0
0.8	0.8	0.8	1.1	1.0	0.3	0.5	0.6	0.6	0.5	0.8
0.7	0.6	1.0	1.8	1.0	1.0	0.2	1.1	0.4	0.4	0.8
2.0	2.3	2.8	0.7	0.8	1.0	1.4	0.4	0.5	0.7	0.7
0.7	1.4	0.9	0.6	1.2	1.7	0.7	0.5	0.8	0.8	0.7
0.8	0.8	0.9	0.8	0.8	0.8	1.3	0.4	1.7	1.4	2.3
1.6	1.4	0.5	0.3	0.7	1.1	1.2	0.7	0.5	2.5	2.1
2.5	1.2	0.8	0.7	0.5	0.7	0.5	0.5	0.9	1.6	2.5
0.1	0.0	0.4	0.6	0.6	0.3	0.9	0.9	0.3	1.0	2.6

0.0	0.0	0.0	0.0	0.3	0.8	0.3	0.0	0.0	0.0	0.0
0.0	0.0	0.1	0.4	0.8	0.2	0.3	0.2	0.0	0.0	0.0
0.0	0.0	0.1	1.4	0.9	0.7	0.1	0.2	0.1	0.0	0.0
0.2	0.7	1.3	0.7	0.7	0.9	1.1	0.2	0.1	0.1	0.0
0.1	0.8	0.7	0.5	1.0	1.5	0.5	0.2	0.1	0.1	0.0
0.1	0.5	0.7	0.7	0.8	0.6	1.1	0.3	0.4	0.5	0.4
0.3	0.7	0.5	0.3	0.6	1.1	0.9	0.6	0.4	0.9	0.9
0.9	0.9	0.8	0.7	0.4	0.7	0.3	0.4	0.6	0.8	1.2
0.1	0.0	0.4	0.4	0.5	0.3	0.8	0.8	0.3	0.8	0.5

Figure 3.6. PSNR gains per macroblock for *Silent Voice* when using the original sequence (left) and when quantizing the reference picture using $\tilde{Q} = 4$ (right).

3.2.5 RELATIONSHIP TO OTHER PREDICTION METHODS

As illustrated, long-term memory MCP benefits from various effects that are quite likely to occur in natural camera-view image sequences. For some of these effects, researchers have proposed alternative methods to exploit them. For instance, short-term frame memory/long-term frame memory prediction [ISO96a] has been proposed to exploit scene cuts. Long-term memory MCP includes this method as a special case when using the *Adaptive Buffering* memory control as shown for the surveillance sequence.

For background memory prediction, researchers have proposed to estimate an additional reference frame for motion compensation that contains the background [MK85, Hep90, Yua93, DM96, ZK98]. For aliasing prediction, a super-resolution image for sub-pixel motion compensation is estimated [Wed99]. The estimation for background memory and aliasing prediction is based on past decoded frames and transmitted parameters since encoder and decoder have to conduct it simultaneously. Therefore, the possible prediction gain highly depends on the accuracy of these estimates.

Additionally, each of the methods (short-term frame memory/long-term frame memory, background memory and aliasing prediction) can only exploit the particular effect it is designed for. When several of these effects occur, a combination of the schemes could be interesting. However, the long-term memory approach can elegantly exploit all of the effects jointly with one simple concept. It can also exploit other long-term statistical dependencies that are not captured by heuristic models.

Hence, it might be more appropriate to view MCP as a statistical optimization problem similar to entropy-constrained vector quantization (ECVQ) [CLG89]. The image blocks to be encoded are quantized using their own code books that consist of image blocks of the same size in the previously decoded frames: the motion search range. A code book entry is addressed by the translational motion parameters which are entropy-coded. The criterion for the block motion estimation is the minimization of a Lagrangian cost function wherein the distortion represented by the prediction error, is weighted against the rate associated with the translational motion parameters using a Lagrange multiplier. The Lagrange multiplier imposes the rate constraint as for ECVQ, and its value directly controls the rate-distortion trade-off [CLG89, SG88, Gir94].

Following this interpretation, the parameters of the ECVQ problem are investigated in the next sections. In Section 3.3, the code book size and its statistical properties are analyzed. The entropy coding is investigated during the course of integrating long-term memory MCP into ITU-T Recommendation H.263 in Section 3.4.

3.3 STATISTICAL MODEL FOR THE PREDICTION GAIN

In this section, the gains that are achievable by long-term memory MCP are statistically modeled. The aim is to arrive at an expression that indicates how a particular strategy for code book adaptation, i.e., search space adaptation, affects the prediction gain. This is important, for example, for a transmission scenario with a small end-to-end delay, where it is not possible to decide whether or not a particular block should be kept or removed from the search space by evaluating future data. Moreover, in Chapter 5, some results of the analysis in this section are exploited providing very significant reductions in computation time.

The analysis starts with distortion values D_m that correspond to the best matches for a block in terms of minimum MSE that is found on the frame with index m and M being the number of reference frames in the multi-frame buffer. Here, it is assumed that the *Sliding Window* memory control is used so that a larger frame index m corresponds to a larger time interval between the current and the reference frame. The matching is conducted for blocks of size 16×16 pixels. The minimization considers ± 16 pixel positions spatially displaced blocks with successive half-pixel refinement.

Figure 3.7 shows the normalized histogram of the measured logarithmic distortion values found on the prior frame $m = 1$ for the set of test sequences in Tab. A.1. The logarithmic distortion L_m for a sequence as a function of the measured MSE values D_m is defined as follows

$$L_m = 10 \log_{10} D_m, \tag{3.1}$$

where m refers to the picture reference parameter. The reason for preferring L_m over D_m is that the resulting probability density function (PDF) is more similar to a Gaussian for which the following computations can be treated analytically. (Please note that the likelihood that $D_m = 0$ and $L_m \to \infty$ is found to be very small in practice.) In Fig. 3.7, a Gaussian PDF is superimposed which is parameterized using the mean and the variance estimates of the measured logarithmic distortion values.

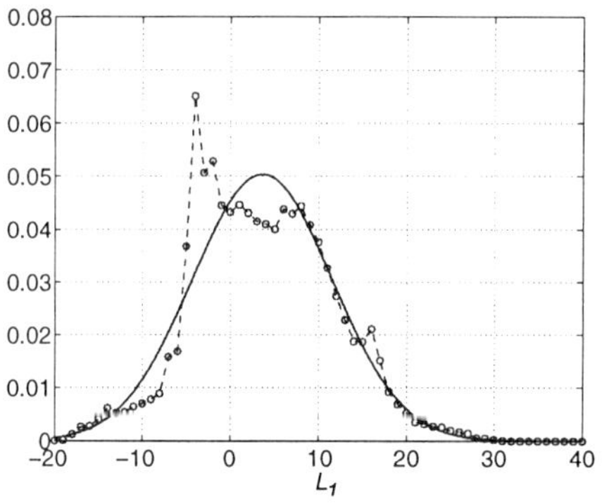

Figure 3.7. Histogram of logarithmic distortions and Gaussian PDF that is adapted via estimating mean and variance given the measured logarithmic distortion values L_1.

The block matching is considered as a random experiment. The vector valued random variable $\mathcal{X}^M$ that is denoted as

$$\mathcal{X}^M = (\mathcal{X}_1 \dots \mathcal{X}_m \dots \mathcal{X}_M)^T \tag{3.2}$$

assigns a vector of M numbers to the outcome of this experiment, which corresponds to the logarithmic distortion values L_m that are found for each of the M

reference frames. The idea is to parameterize a joint PDF $f_{\mathcal{X}^M}$ that describes the probability of the vector-valued outcome of the random experiment. Then, the minimization is computed analytically for the model PDF. This analytical result is compared to the actual minimization result to validate the accuracy of the model and the conclusions drawn from it.

Measurements show that the distortion values that correspond to the M reference frames for each block are correlated. Hence, a correlated vector-valued random variable has to be considered. The PDF, that describes the random experiment, is assumed to be a jointly Gaussian of the form

$$f_{\mathcal{X}^M}(\boldsymbol{x}) = \frac{1}{(2\pi)^{M/2}|\boldsymbol{C}|^{1/2}} e^{-\frac{1}{2}(\boldsymbol{x}-\boldsymbol{\mu})^T \boldsymbol{C}^{-1}(\boldsymbol{x}-\boldsymbol{\mu})} \tag{3.3}$$

with

$$\boldsymbol{C} = \begin{pmatrix} c_{1,1} & c_{1,2} & \cdots & c_{1,M} \\ c_{2,1} & c_{2,2} & \cdots & c_{2,M} \\ \vdots & \vdots & & \vdots \\ c_{M,1} & c_{M,2} & \cdots & c_{M,M} \end{pmatrix} \quad \text{and} \quad \boldsymbol{\mu} = \begin{pmatrix} \mu_1 \\ \mu_2 \\ \vdots \\ \mu_M \end{pmatrix} \tag{3.4}$$

being covariance matrix and mean vector, respectively. The following assumptions are made to further simplify the model

$$c_{n,m} = \begin{cases} \sigma^2 & n = m \\ \rho \cdot \sigma^2 & n \neq m \end{cases} \quad \text{and} \quad \mu_1 = \mu_2 = \cdots = \mu_M = \mu. \tag{3.5}$$

In the following, the jointly Gaussian PDF in (3.3) with the assumptions in (3.5) of M random variables with mean μ, variance σ, and correlation factor ρ is denoted as $\mathcal{N}(\boldsymbol{x}, M, \mu, \sigma, \rho)$.

The minimization is conducted by drawing an M-tuple $\mathcal{X}^M = (\mathcal{X}_1, \ldots, \mathcal{X}_M)$ and choosing the minimum element. This minimum element is considered as the outcome of another random experiment and the associated random variable is called $\mathcal{Y}_{1,M}$, where the indices of $\mathcal{Y}$ correspond to the first and the last index of the random variable for which the minimization is conducted. As the model parameter that corresponds to the average logarithmic distortion reduction for long-term memory prediction, the difference $\Delta_{1,M}$ of the expected values between $\mathcal{X}_1$ and $\mathcal{Y}_{1,M}$ is considered.

While the numerical minimization is a rather simple operation, its analytical treatment is difficult in case of correlated random variables. But, for the case $M = 2$, the analytical computation of the expected values after minimization is possible as shown in Appendix B. The mean difference $\Delta_{1,2}$ is given as

$$\Delta_{1,2} = E\{\mathcal{X}_1\} - E\{\mathcal{Y}_{1,2}\} = \mu - E\{\mathcal{Y}_{1,2}\} = \sigma\sqrt{\frac{1-\rho}{\pi}}. \tag{3.6}$$

Another quantity that changes after minimization is the variance. Thus the ratio $\xi_{1,2}^2$ between the variances after and before minimization is given as

$$\xi_{1,2}^2 = \frac{E\left\{\mathcal{Y}_{1,2}^2\right\} - E\left\{\mathcal{Y}_{1,2}\right\}^2}{E\left\{\mathcal{X}_1^2\right\} - E\left\{\mathcal{X}_1\right\}^2} = \frac{E\left\{\mathcal{Y}_{1,2}^2\right\} - E\left\{\mathcal{Y}_{1,2}\right\}^2}{\sigma^2} = 1 - \frac{1-\rho}{\pi}. \tag{3.7}$$

Hence, in order to minimize $M = 2^K$ random variables, a cascade of K dyadic minimizations of two random variables is utilized as illustrated in Fig. 3.8. This approach implicitly assumes that the result of the minimization of two jointly Gaussian random variables is approximately a Gaussian random variable as well. The validity of this assumption will be verified later.

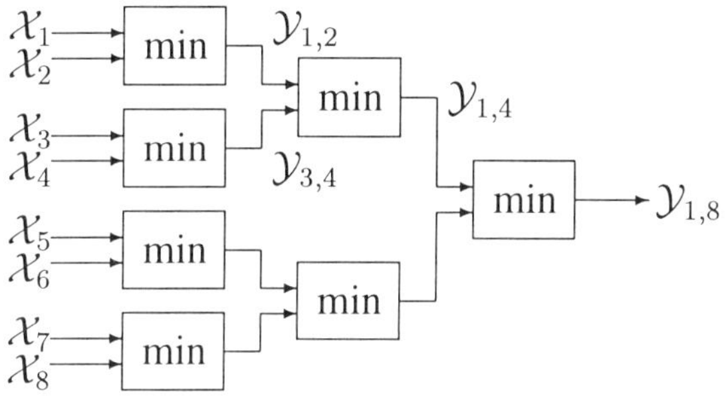

Figure 3.8. Cascaded minimization of random variables.

Let us consider the case $K = 2$, i.e., $M = 4$. The random variable $\mathcal{Y}_{1,2}$ is assumed to be jointly Gaussian distributed of the form

$$\mathcal{N}(\boldsymbol{x}, 2, \mu_{1,2}, \sigma_{1,2}, \rho_{1,2}) \tag{3.8}$$

with

$$\begin{aligned} \mu_{1,2} &= \mu - \Delta_{1,2} = \mu - \sigma\sqrt{\frac{1-\rho}{\pi}}, \\ \sigma_{1,2} &= \sigma\xi_{1,2} = \sigma\sqrt{1 - \frac{1-\rho}{\pi}}, \\ \rho_{1,2} &= \rho\xi_{1,2}^{-2} = \rho\left(1 - \frac{1-\rho}{\pi}\right)^{-1}. \end{aligned} \tag{3.9}$$

The scaling of $\rho_{1,2}$ is needed because of the definition of it as the ratio between covariance and variance. The same distribution is assumed for $\mathcal{Y}_{3,4}$. The random variables $\mathcal{Y}_{1,2}$ and $\mathcal{Y}_{3,4}$ are fed into a minimization yielding the random variable $\mathcal{Y}_{1,4}$. The mean, variance and correlation factor of $\mathcal{Y}_{1,4}$ can then be computed via (3.6) and (3.7). The repeated application of this procedure yields a geometric series for which a closed form expression exists. Hence, the mean difference for $M = 2^K$ random variables is approximately given as

$$\Delta_{1,M} \approx \sigma\sqrt{\frac{1-\rho}{\pi}}\cdot\frac{1-\alpha^K}{1-\alpha} = \sigma\sqrt{\frac{1-\rho}{\pi}}\cdot\frac{1-\alpha^{\log_2 M}}{1-\alpha}, \quad \text{with} \quad \alpha = \sqrt{1-\frac{1}{\pi}}. \tag{3.10}$$

In Figure 3.9, the result of the cascaded minimization for various values of the correlation parameter ρ is shown for a jointly normal random variable with $\sigma = 1$. As a comparison to the prediction in (3.10) that is depicted with solid lines, the circles show the result of a numerical minimization of data that are drawn from a jointly Gaussian distribution.

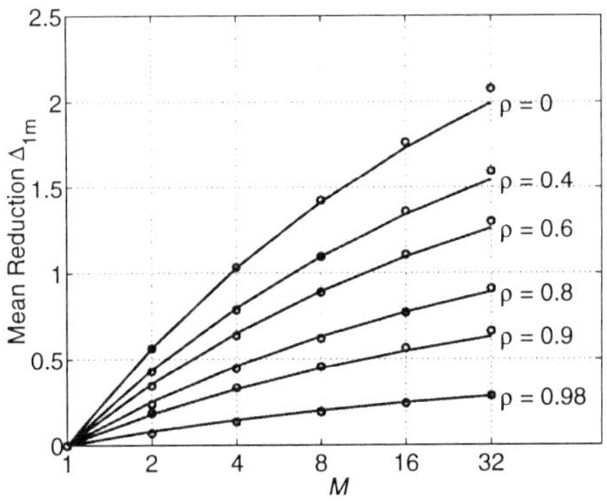

Figure 3.9. Result of the cascaded minimization using (3.10).

In order to relate the model prediction of (3.10) to long-term memory prediction, various experiments are conducted. For that, the sequences in the test set and the conditions in Tab. A.1 are employed. (The sequence *News* is excluded here since its background is artificial and block matching may result in $D_m = 0$ for some blocks.) The experiment consists of block matching for blocks of size 16×16 or 8×8 for integer-pixel and half-pixel accuracy. For each reference frame, the best block match is determined in terms of MSE when searching each original frame in a range of ± 16 spatially displaced pixels in horizontal and vertical direction. The measured MSE values for each block are mapped into the logarithmic distortion using (3.1). Then, the minimization is conducted over the set of $M = 2, 4, 8, 16$, and 32 reference frames and the resulting average minimum distortion values are subtracted form the average distortion that is measured when referencing the prior frame only. The result of this experiment is shown in Fig. 3.10 using the circles.

The prediction by the statistical model is depicted by the solid lines in Fig. 3.10. This prediction is obtained by estimating the mean, variance and correlation factor for the measured logarithmic distortion values. The mean

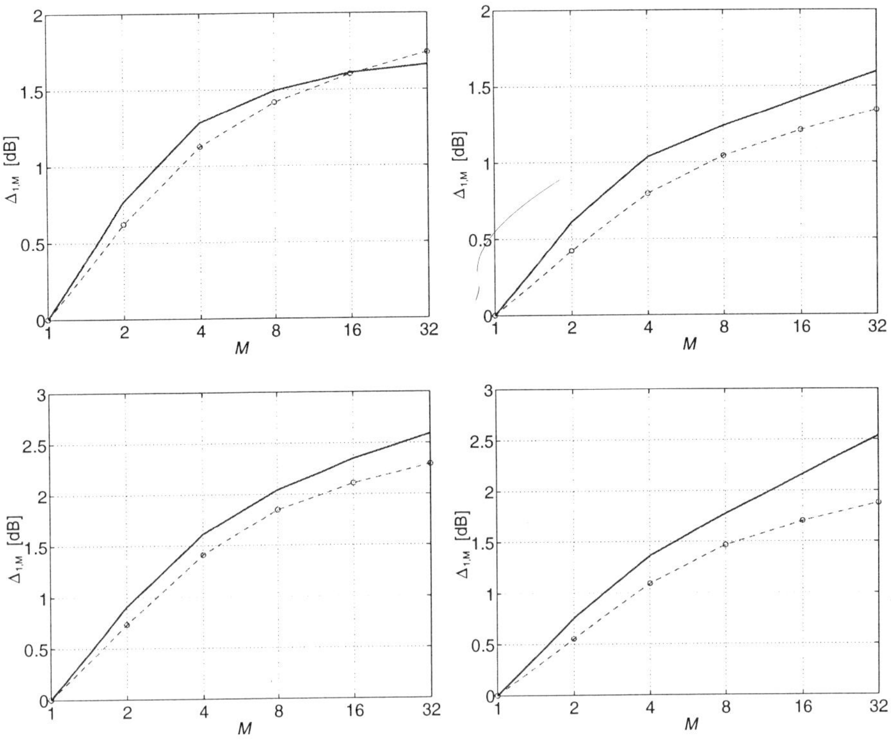

Figure 3.10. Measured and model-predicted logarithmic distortion reduction in dB vs. number of reference frames M. The results are shown for four cases: 16 × 16 blocks and integer-pixel accuracy (top left), 16 × 16 blocks and half-pixel accuracy (top right), 8 × 8 blocks and integer-pixel accuracy (bottom left), 8 × 8 blocks and half-pixel accuracy (bottom right).

values, variances, and correlation factors depend on the time interval between the current and the reference picture. However, the analysis of the minimization using the jointly Gaussian PDF is conducted for identical mean values, variances and correlation factors in (3.4) for all reference frames. Hence, the measured logarithmic distortion values are permuted before estimating mean values, variances and correlation factors. Assume N to be the number of blocks in the set of considered sequences. Further, assume the distortion values being gathered in a $N \times M$ matrix with the columns corresponding to reference frames $1 \ldots M$ and the entries in each row relate to a block. Permuting means, that the columns are randomly shuffled for each row in order to achieve equal estimates over the columns. Then given these randomly shuffled matrix of data, the correlation factor as well as mean and variance are estimated.

Several observations can be made

- Recognizing that only four estimated values (M, μ, σ, and ρ) are used to predict the distortion reduction, the measured results and the model prediction are fairly close.

- The relative gains for integer-pixel accuracy (left-hand side plots in Fig. 3.10) are larger than for half-pixel accuracy (right-hand side plots in Fig. 3.10). The aliasing compensation effect and the corresponding subpixel position in the past are an explanation for this effect. Statistically, the difference between the measured mean values μ for the two cases get smaller as M increases.

- The relative gains for 16×16 blocks (upper two plots in Fig. 3.10) are smaller than for 8×8 blocks (lower two plots in Fig. 3.10). Statistically, the larger gains are due to larger values of σ and because the average logarithmic distortion μ increases faster for 16×16 blocks as for 8×8 blocks when M increases. In general, it becomes more likely to find a good match for small blocks than for large blocks as the time interval between the frames increases.

The prediction of the logarithmic distortion reduction by the statistical model is dependent on four variables M, μ, σ, and ρ. Increasing M always provides a lower MSE. For the considered range of $2 \leq M \leq 32$ reference frames, the mean difference in (3.10) can be approximated by

$$\Delta_{1,M} \approx \sigma \sqrt{\frac{1-\rho}{\pi}} (\log_2 \log_2 M + 1) \tag{3.11}$$

which shows that the mean difference in dB is proportional to the log-log of the number of reference frames. The mean μ is mainly influenced by the probability of finding a good match in frames that are several time instants away from the current frame. This probability is much larger as the block size decreases and therefore the gains when employing more reference frames increase for decreasing block size. The variance σ and the correlation factor ρ play an important role. These parameters specify the slope of the distortion reduction given the number of reference frames. By decreasing the correlation factor, the logarithmic distortion reductions are larger. This suggests a buffering rule in which blocks that are "too similar" are rejected because they lead to large values of ρ. The application of this rule provides very significant reductions in computation time at minor losses in rate-distortion performance as demonstrated in Chapter 5.

3.4 INTEGRATION INTO ITU-T RECOMMENDATION H.263

In the previous sections, it has been shown that long-term memory MCP can provide a significant MSE reduction, when considering the prediction error.

In this section, it is demonstrated that long-term memory MCP also yields improved rate-distortion performance when being integrated into a hybrid video coder, where the side information for the picture reference parameter has to be considered. In the following, the rate-distortion trade-off for long-term memory MCP is analyzed followed by a presentation of the rate-distortion performance of the complete codec.

3.4.1 RATE-CONSTRAINED LONG-TERM MEMORY PREDICTION

The motion vector $\boldsymbol{m}_i$ to predict a block $\boldsymbol{S}_i$ has to be transmitted as side information requiring additional bit-rate. Given M reference frames, the Lagrangian cost function in (2.8) is minimized for motion estimation. Typically, the set of positions in the search space in horizontal and vertical direction and over the reference frames is given as

$$\mathcal{M} = [-16\ldots16] \times [-16\ldots16] \times [1\ldots M]. \tag{3.12}$$

The distortion $D_{\text{DFD}}(\boldsymbol{S}_i, \boldsymbol{m})$ is computed either using SSD or SAD, while $R_{\text{MOTION}}(\boldsymbol{S}_i, \boldsymbol{m})$ is given by the bits for the spatial displacements and the picture reference parameter. The motion search first determines the best integer-pixel accurate motion vector. Then, the final motion vector $\boldsymbol{m}_i$ is determined by minimizing (2.8) when searching the eight half-pixel positions that surround the integer-pixel accurate motion vector.

A trade-off between prediction gain and motion bits can be achieved by controlling λ_{MOTION}. Figure 3.11 shows the result of a MCP experiment that is conducted to illustrate that trade-off. The experiment consists of two steps:

1. The TMN-10 coder is run employing quantizer values $Q_{\text{REF}} \in \{4, 10, 25\}$. The resulting bit-streams are decoded and the reconstructed frames are employed as reference frames in the next step.

2. Given the coded reference frames, the MCP signal is computed. Similar to H.263+, the coder has the option to represent each 16×16 block either using one motion vector or four motion vectors. In the latter case, each motion vector corresponds to an 8×8 block. The motion estimation is conducted by minimizing (2.8) for both block sizes separately when employing the SSD distortion measure. Then, given the Lagrangian costs for one motion vector and the sum of the Lagrangian costs for the four motion vectors, the decision is made between the two options again by choosing the minimum [SB91]. For each macroblock, first one bit is transmitted that indicates whether the macroblock region is represented as a copy of the macroblock in the same location in the prior decoded picture. If the macroblock is not copied, then another bit is transmitted that indicates whether motion

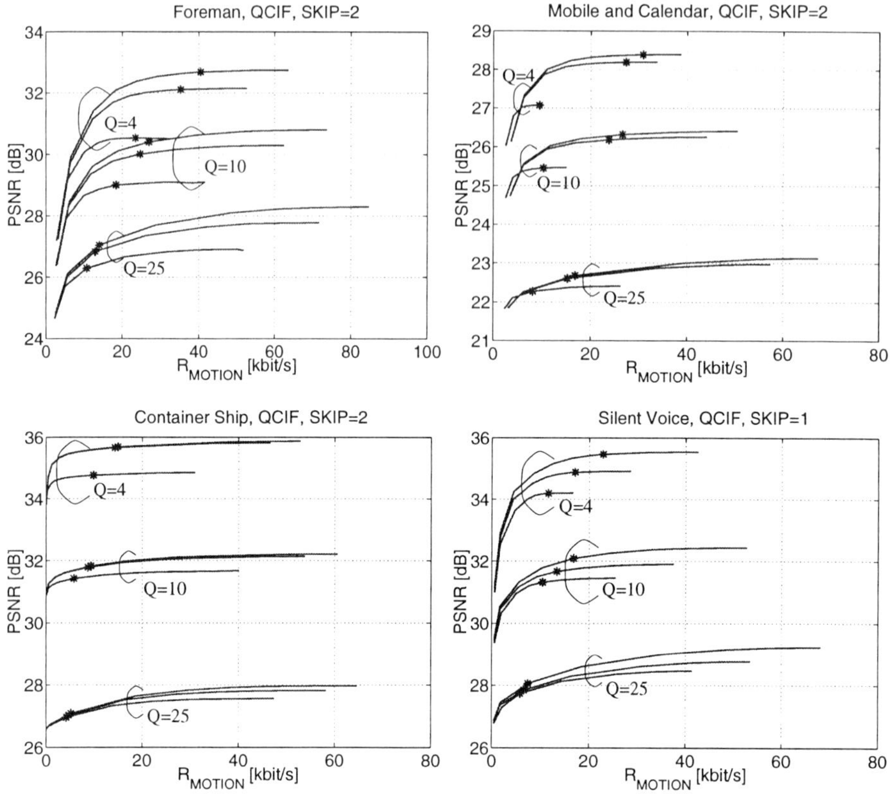

Figure 3.11. PSNR of motion-compensated frames vs. motion bit-rate.

compensation is conducted using 16×16 or 8×8 blocks. Dependent on this choice, either one or four spatial displacements and picture reference parameters are transmitted. The motion search proceeds over the range in (3.12) with $M \in \{1, 10, 50\}$.

In Fig. 3.11, the PSNR in dB between the motion-compensated frames and the corresponding original frames is depicted vs. bit-rate for the motion vectors measured in kbit/s. As marked in the pictures, the curves correspond to three settings of the DCT quantizer value for the reference frames which are $Q_{\text{REF}} = 4$, 10, and 25. For each quantizer value Q_{REF}, three curves are depicted that correspond to MCP using $M = 1, 10$, and 50 reference frames. A larger number of reference frames always means a larger PSNR value for the case $\lambda_{\text{MOTION}} = 0$ which is the point of maximum motion bit-rate for each curve. Each curve is generated by varying the Lagrange parameter λ_{MOTION} when minimizing (2.8). Several observations can be made:

- The prediction gains in terms of PSNR due to an increased number of reference frames are reduced as the DCT quantizer value of the reference frames increases.

- The motion bit-rate increases significantly for $\lambda_{\text{MOTION}} = 0$ as the distortion in the reference frames and their number M increases. This shows the importance of the rate constraint for motion estimation.

- The points that are marked by stars correspond to the case where the choice $\lambda_{\text{MOTION}} = 0.85 Q_{\text{REF}}^2$ is made. These points seem to be good compromises between prediction performance and motion bit-rate. This is because, the additional bit-rate for the cases $M > 1$ compared to the point for $M = 1$ decreases as the DCT quantizer and with that the slope of the rate-distortion curve becomes larger (see Section 2.4 for a detailed analysis).

For the results in Fig. 3.11, the variable length code that is specified in ITU-T Recommendation H.263+ [ITU98a] has been employed for each component of the spatial displacement vector. For the transmission of the picture reference parameter, one variable length code has been generated using an iterative design approach similar to the algorithm for ECVQ in [CLG89]. The indices in this table correspond to those for the multi-frame buffer.

3.4.2 RATE-DISTORTION PERFORMANCE

Figure 3.12 shows the average PSNR from reconstructed frames produced by the TMN-10 codec and the long-term memory prediction codec vs. overall bit-rate. For all cases, the Annexes D, F, I, J, and T of the ITU-T Recommendation H.263+ are enabled [ITU98a]. The size of the long-term memory is selected as 2, 10, and 50 frames and the syntax employed is similar to that of Annex U of H.263++ [ITU00]. The curves are generated by varying the Lagrange parameters and the DCT quantization parameter accordingly when encoding the sequences *Foreman*, *Mobile & Calendar*, *Container Ship*, and *Silent Voice* using the conditions of Tab. A.1. The points marked on the curves correspond to values computed from the entire sequence. The long-term memory buffer is built up simultaneously at encoder and decoder by reconstructed frames. The results are averaged excluding the first 50 frames, in order to avoid the effects at the beginning of the sequence. Please note that the results do not change much when the first 50 frames are considered as well. More important are the statistical characteristics of the sequences.

For the results in Fig. 3.12, the same sequences are used as for the motion compensation experiment in Fig. 3.11. Most of the gains and tendencies observed for the motion compensation experiment carry over to the case when the complete coder is employed. For example, the motion compensation experiment for the sequence *Container Ship* indicates no improvement when increas-

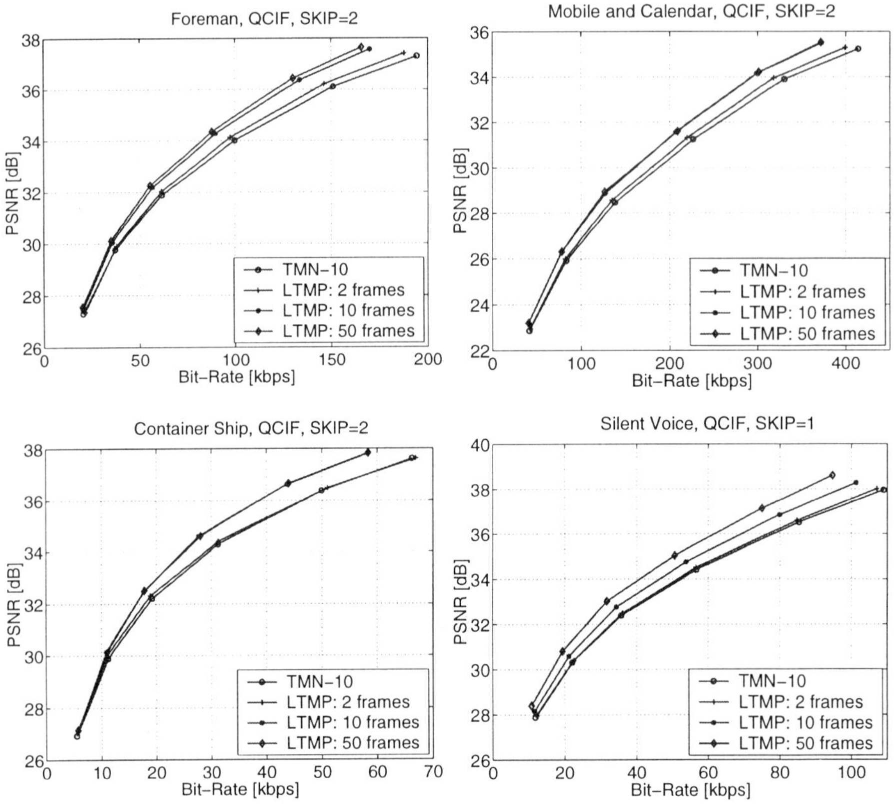

Figure 3.12. PSNR of reconstructed frames vs. overall bit-rate.

ing the memory size from 10 to 50 frames. This observation can also be made in the corresponding plot in Fig. 3.12. On the other hand, a significant relative gain can be obtained for both experiments for the sequence *Silent Voice* when moving from 10 to 50 frames.

The PSNR gains obtained when comparing the long-term memory MCP codec with memory $M = 50$ to the TMN-10 are between 0.9 to 1.5 dB for the four sequences in Fig. 3.12. But for most sequences, a memory size of $M = 10$ frames already provides most of the gain. This can be verified by looking at Fig. 3.13. The left-hand side plot of Fig. 3.13 shows the average bit-rate savings that are measured for each sequence at fixed PSNR values of 32, 34 and 36 dB. For that, rate-distortion curves like the ones in Fig. 3.12 are generated by varying the DCT quantizer and the Lagrange parameter accordingly. The bit-rate corresponds to the overall bit-rate that has to be transmitted to reconstruct each sequence at the decoder and distortion is computed as average PSNR over all frames. The intermediate points of the rate-distortion curves are interpolated and the bit-rate that corresponds to a given PSNR value is obtained. The curves in Fig. 3.13 are obtained via computing the mean of the bit-rate savings for each sequence. This procedure is conducted for all sequences and the plot shows the average of the bit-rate savings for each sequence. The average bit-rate savings are very similar for the three different levels of reproduction quality. When considering 34 dB reproduction quality and employing 10 reference frames, an average bit-rate reduction of 12 % can be observed. When employing 50 reference frames, the bit-rate savings are around 17 %.

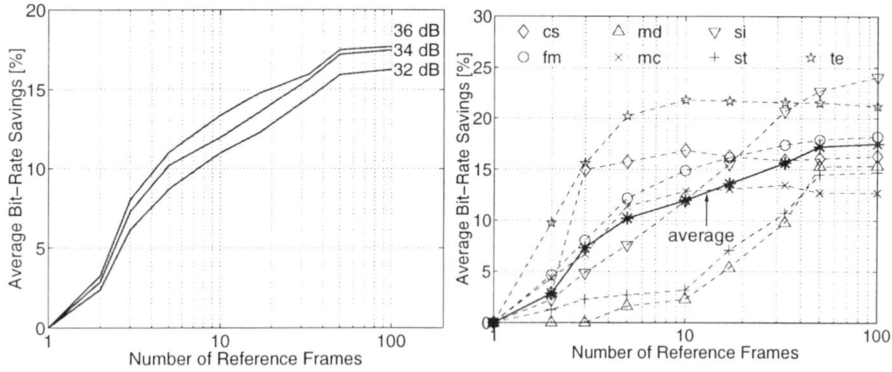

Figure 3.13. Average bit-rate savings vs. number of reference frames.

The right-hand side plot of Fig. 3.13 shows the average bit-rate savings at 34 dB PSNR for the set of test sequences where the result for each sequence is depicted using dashed lines. The abbreviations *fm, mc, st, te, cs, md,* and *si* correspond to those in Tab. A.1. The result for the sequence *News* will be shown later. The bit-rate reductions differ quite significantly among the various sequences. For sequences with uncovered background effects, like *Container Ship* and *Tempete*, most of the gain is obtained when using only 3 or 5 reference frames. Other sequences like *Stefan* and *Mother & Daughter* seem to pick up when the memory size increases to 50 frames, while the gains at 10 frames are rather small.

The exceptional bit-rate savings for the sequence *News* should be mentioned that have already been indicated when comparing the prediction performance in Fig. 3.4. In Fig. 3.4, the repetition of the dancers in the background of the scene provides a PSNR gain up to 20 dB for the corresponding part of the image. The result for the coding experiment is shown in Fig. 3.14. The same settings as for the results in Fig. 3.12 have been employed. The PSNR gains for memory $M = 50$ compared to the TMN-10 coder are more than 6 dB or correspond to bit-rate savings of more than 60 %.

In order to give a further confirmation of the performance of long-term memory MCP, the coder has been run on 8 self-recorded natural sequences. These sequences show typical interactive video phone contents. Also for these se-

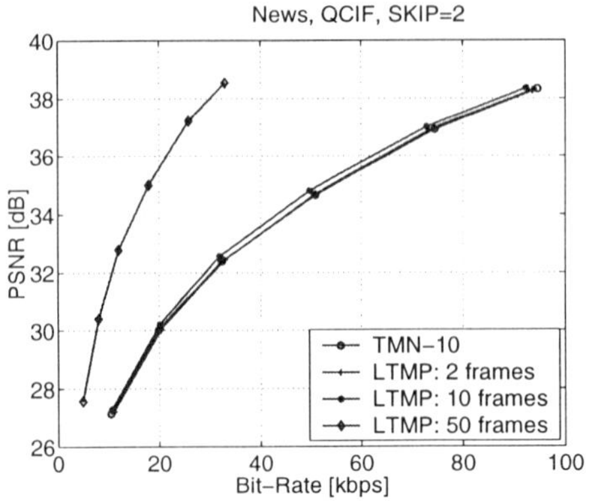

Figure 3.14. PSNR vs. overall bit-rate for the sequence *News*. Simulation conditions as for Fig. 3.12.

quences average bit-rate savings between 12.5 % and 30 % have been obtained when using $M = 50$ reference frames.

3.5 DISCUSSION AND OUTLOOK

The long-term memory video compression architecture and the rate-constrained coder control can serve as a very general approach to improve MCP. In general, any technique that provides useful image data for MCP may be utilized to generate reference frames. These techniques may include Sprites [DM96], "layers" from the Layered Coding scheme [WA94], or Video Object Planes (VOPs) as defined within MPEG-4 [ISO98b]. The decoder just needs to be informed about parameters that are employed to simultaneously generate the reference frames and be given a reference coordinate system to conduct the motion compensation. Based on rate-distortion efficiency, the encoder has to decide whether or not to include a particular frame. Generating frames by one of the techniques mentioned requires additional computation. Also, the sequences have to lend themselves to representations with Sprites, layers, or VOPs. In Chapter 4, such an extension of the long-term memory concept is presented, where reference frames are generated by affine warping of previous decoded frames.

Another approach to enhance MCP is to combine multi-hypothesis prediction as described in Section 1.5.3 and the long-term memory approach as published in [FWG98, WFG98, FWG00b, FWG00a]. The idea is to employ the B-frame concept while referencing only decoded frames in the temporal past. This way, the delay problem does not occur as for B-frames. The estimation of the two motion vectors is conducted using an iterative approach to minimize a Lagrangian cost function. As reported in [FWG00b, FWG00a], the combination of the multi-hypothesis and the long-term memory approaches yields more than additive gains. For the sequence *Mobile & Calendar*, a bit-rate saving of 10 % is obtained for the multi-hypothesis codec when employing $M = 1$ reference picture in comparison to TMN-10. Long-term memory MCP using $M = 10$ reference frames provides a bit-rate reduction of 13 % against TMN-10 while for the combined coder a bit-rate reduction of 32 % is reported. Similarly, for the sequence *Foreman*, a bit-rate reduction of 23 % is obtained against TMN-10 for the combined multi-hypothesis and long-term memory codec.

3.6 CHAPTER SUMMARY

A new technique for MCP is presented that exploits long-term statistical dependencies in video sequences. These dependencies include scene cuts, uncovered background, and high-resolution texture with aliasing. For those dependencies, other researchers have proposed alternative methods for their exploitation. However, these methods rely on the occurrence of the particular effect they are designed for. The new technique, long-term memory MCP, exploits all these effects simultaneously by one single concept.

A statistical model for the prediction gain is developed. The statistical model as well as measurements show that an increasing the number of reference pictures always provides improved prediction gains. The prediction gains measured as PSNR in dB are roughly proportional to the log-log of the number of reference frames. The analysis yields the result that extending the search space by blocks which are "too similar" to the existing blocks only yields small prediction gains. This suggests a buffering rule in which blocks that are "too similar" are rejected resulting in drastic reductions in computation time as shown in Chapter 5.

The integration of long-term memory MCP into an H.263-based hybrid video codec shows that the bit-rate overhead which is incurred by the picture reference parameter is well compensated by the prediction gain. When considering 34 dB reproduction quality and employing 10 reference frames, average bit-rate savings of 12 % against TMN-10 can be observed for the set of test sequences. When employing 50 reference frames, the average bit-rate savings against TMN-10 are 17 % and the minimal bit-rate savings inside the test set are 13 % while the maximal bit-rate savings are reported to be up to 23 %. These average bit-rate savings relate to PSNR gains between 0.7 to 1.8 dB. For some image sequences, very significant bit-rate savings of more than 60 % can be achieved.

The ideas and results presented in this chapter lead to the creation of an extension of ITU-T Recommendation H.263 via adopting the feature as Annex U to H.263++ [ITU00]. Moreover, the currently ongoing H.26L project of the ITU-T Video Coding Experts Group contains long-term memory MCP as an integral part of the codec design.

Chapter 4

AFFINE MULTI-FRAME MOTION-COMPENSATED PREDICTION

While long-term memory MCP extends the motion model to exploit long-term dependencies in the video sequence, the motion model remains translational. Independently moving objects in combination with camera motion and focal length change lead to a sophisticated motion vector field which may not be efficiently approximated by a translational motion model. With an increasing time interval between video frames as is the case when employing long-term memory MCP, this effect is further enhanced since more sophisticated motion is likely to occur.

In this chapter, long-term memory MCP is combined with affine motion compensation. Several researchers have approached the control of an affine motion coder as an optimization problem where image segmentation and affine motion parameter estimation have to be conducted jointly for rate-distortion efficient results [San91, YMO95, CAS$^+$96, FVC87, HW98]. The methodology in this work differs from previous approaches in that the joint optimization problem is circumvented by employing an approach that is similar to global motion compensation [Höt89, JKS$^+$97, ISO97a].

The idea is to determine several affine motion parameter sets on sub-areas of the image. Then, for each affine motion parameter set, a complete reference frame is warped and inserted into the multi-frame buffer. Given the multi-frame buffer of decoded frames and affine warped versions thereof, block-based translational MCP and Lagrangian coder control are utilized as described in Chapter 3. The affine motion parameters are transmitted as side information requiring additional bit-rate. Hence, the utility of each reference frame and with that each affine motion parameter set is tested for its rate-distortion efficiency.

In Section 4.1, the extension of long-term memory MCP to affine motion compensation is explained. The coder control is described in Section 4.2, where the estimation procedure for the affine motion parameters and the reference picture warping are presented. Then, the determination of the efficient number of affine motion parameter sets is described. Finally, in Section 4.3, experimental results are presented that illustrate the improved rate-distortion performance in comparison to TMN-10 and long-term memory MCP.

4.1 AFFINE MULTI-FRAME MOTION COMPENSATION

In this section, the structure of the affine multi-frame motion compensation is explained. First, the extension of the multi-frame buffer by warped versions of decoded frames is described. Then, the necessary syntax extensions are outlined and the affine motion model, i.e., the equations that relate the affine motion parameters to the pixel-wise motion vector field are presented.

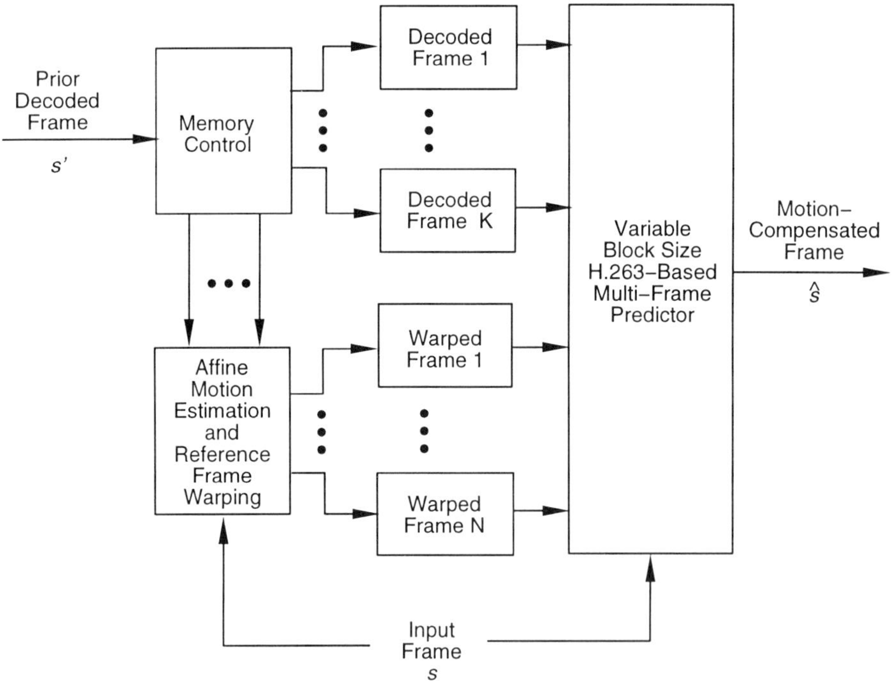

Figure 4.1. Block diagram of the affine multi-frame motion-compensated predictor.

The block diagram of the multi-frame affine motion-compensated predictor is depicted in Fig. 4.1. The motion-compensated predictor utilizes $M = K+N$ ($M \geq 1$) picture memories. The M picture memories are composed of two sets:

1. K past decoded frames and

2. N warped versions of past decoded frames.

The H.263-based multi-frame predictor conducts block-based MCP using all $M = K + N$ frames and produces a motion-compensated frame. This motion-compensated frame is then used in a standard hybrid DCT video coder [ITU98a, SW98]. The N warped reference frames are determined using the following two steps:

1. Estimation of N affine motion parameter sets between the K previous frames and the current frame.

2. Affine warping of N reference frames.

The number of efficient reference frames $M^* \leq M$ is determined by evaluating their rate-distortion efficiency in terms of Lagrangian costs for each reference frame. The M^* chosen reference frames with the associated affine motion parameter sets are transmitted in the header of each picture. The order of their transmission provides an index that is used to specify a particular reference frame on the block basis. The decoder maintains only the K decoded reference frames and does not have to warp N complete frames for motion compensation. Rather, for each block or macroblock that is compensated using affine motion compensation, the translational motion vector and the affine motion parameter set are combined to obtain the displacement field for that image segment.

Figures 4.2 and 4.3 show an example for affine multi-frame prediction. The left-hand side in Fig. 4.2 is the most recent decoded frame that would be the only frame to predict the right-hand side in Fig. 4.2 in standard-based video compression. Four out of the set of additionally employed reference frames are shown in Fig. 4.3. Instead of just searching over the previous decoded frame (Fig. 4.2a), the block-based motion estimator can also search positions in the additional reference frames like the ones depicted in Fig. 4.3 and transmits the corresponding spatial displacement and picture reference parameter.

4.1.1 SYNTAX OF THE VIDEO CODEC

In a well-designed video codec, the most efficient concepts should be combined in such a way that their utility can be adapted to the source signal without significant bit-rate overhead. Hence, the proposed video codec enables the utilization of variable block-size coding, long-term memory prediction and affine motion compensation using such an adaptive method, where the use of the multiple reference frames and affine motion parameter sets can be signaled with very little overhead.

The parameters for the chosen reference frames are transmitted in the header of each picture. First, their actual number M^* is signaled using a variable length code. Then, for each of the M^* reference frames, an index identifying one of the past K decoded pictures is transmitted. This approach is similar to

(a) (b)

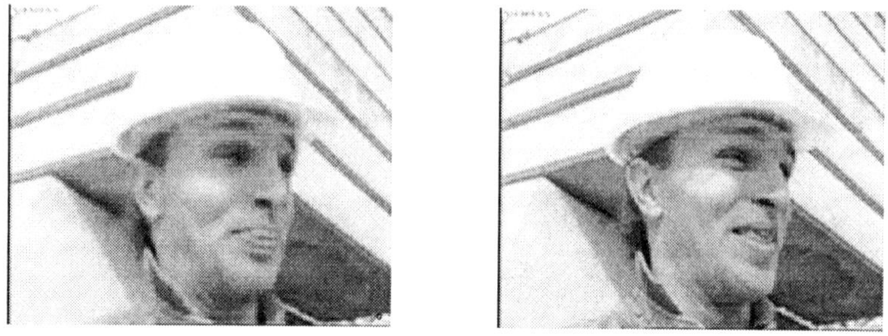

Figure 4.2. Two frames from the QCIF test sequence *Foreman*, (a): previous decoded frame, (b): original frame.

Figure 4.3. Four additional reference frames. The upper left frame is a decoded frame that was transmitted 2 frame intervals before the previous decoded frame. The upper right frame is a warped version of the decoded frame that was transmitted 1 frame interval before the previous frame. The lower two frames are warped versions of the previous decoded frame.

the *Index Mapping* memory control in Chapter 3. This index is followed by a bit signaling whether the indicated decoded frame is warped or not. If that bit indicates a warped frame, the corresponding six affine motion parameters are transmitted. This syntax allows the adaptation of the multi-frame affine coder to the source signal on a frame-by-frame basis without incurring much overhead. Hence, if affine motion compensation is not efficient, one bit is enough to turn it off.

4.1.2 AFFINE MOTION MODEL

In this work an affine motion model is employed that describes the relationship between the motion of planar objects and the observable motion field in the image plane via a parametric expression. This model can describe motion such as translation, rotation, and zoom using six parameters $\boldsymbol{a} = (a_1, a_2, a_3, a_4, a_5, a_6)^T$. For estimation and transmission of the affine motion parameter sets, the orthogonalization approach in [KNH97] is adopted. The orthogonalized affine model is used to code the displacement field $(m_x[\boldsymbol{a}, x, y], m_y[\boldsymbol{a}, x, y])^T$ and to transmit the affine motion parameters using uniform scalar quantization and variable length codes. In [KNH97] a comparison was made to other approaches indicating the efficiency of the orthogonalized motion model. The motion model used for the investigations in this chapter is given as

$$\begin{aligned} m_x[\boldsymbol{a}, x, y] &= \frac{w-1}{2}\left[a_1 c_1 + a_2 c_2\left(x - \frac{w-1}{2}\right) + a_3 c_3\left(y - \frac{h-1}{2}\right)\right], \\ m_y[\boldsymbol{a}, x, y] &= \frac{h-1}{2}\left[a_4 c_1 + a_5 c_2\left(x - \frac{w-1}{2}\right) + a_6 c_3\left(y - \frac{h-1}{2}\right)\right]. \end{aligned} \tag{4.1}$$

in which x and y are discrete pixel locations in the image with $0 \leq x < w$ and $0 \leq y < h$ and w as well as h being image width and height. The coefficients c_1, c_2, and c_3 in (4.1) are given as

$$\begin{aligned} c_1 &= \frac{1}{\sqrt{w \cdot h}}, \\ c_2 &= \sqrt{\frac{12}{w \cdot h \cdot (w-1) \cdot (w+1)}}, \\ c_3 &= \sqrt{\frac{12}{w \cdot h \cdot (h-1) \cdot (h+1)}}. \end{aligned} \tag{4.2}$$

The affine motion parameters a_i are quantized as follows

$$\tilde{a}_i = \frac{Q(\Delta \cdot a_i)}{\Delta} \quad \text{and} \quad \Delta = 2, \tag{4.3}$$

where $Q(\cdot)$ means rounding to the nearest integer value. The quantization levels of the affine motion parameters $q_i = \Delta \cdot \tilde{a}_i$ are entropy-coded and transmitted.

It has been found experimentally that similar coding results are obtained when varying the coarseness of the motion coefficient quantizer Δ in (4.3) from 2 to 10. Values of Δ outside this range, i.e., larger than 10 or smaller than 2, adversely affect coding performance. Typically, an affine motion parameter set requires between 8 and 40 bits for transmission rate.

4.2 RATE-CONSTRAINED CODER CONTROL

In the previous section, the video architecture and syntax are described. Ideally, the coder control should determine the coding parameters so as to achieve a rate-distortion efficient representation of the video signal. This problem is compounded by the fact that typical video sequences contain widely varying content and motion, that can be more effectively quantized if different strategies are permitted to code different regions. For the affine motion coder, one additionally faces the problem that the number of reference frames has to be determined since each warped reference frame is associated to an overhead bit-rate. Therefore, the affine motion parameter sets must be assigned to large image segments to keep their number small. In most cases however, these large image segments usually cannot be chosen so as to partition the image uniformly. The proposed solution to this problem is as follows:

A. Estimate N affine motion parameter sets between the current and the K previous frames each corresponding to one of N initial clusters.

B. Generate the multi-frame buffer which is composed of K past decoded frames and N warped frames that correspond to the N affine motion parameter sets.

C. Conduct multi-frame block-based hybrid video encoding on the $M = N+K$ reference frames.

D. Determine the number of affine motion parameter sets that are efficient in terms of rate-distortion performance.

In the following, steps A-D are described in detail.

4.2.1 AFFINE MOTION PARAMETER ESTIMATION

A natural camera-view scene may contain multiple independently moving objects in combination with camera motion and focal length change. Hence, region-based coding attempts to separate the effects via a scene segmentation and successive coding of the resulting image segments. In this work, an explicit segmentation of the scene is avoided. Instead, the image is partitioned into blocks of fixed size which are referred to as clusters in the following. For each cluster one affine motion parameter set is estimated that describes the motion inside this cluster between a decoded frame and the current original frame.

The estimation of the affine motion parameter set for each cluster is conducted in four steps:

1. Estimation of L translational motion vectors as initialization to the affine refinement.

2. Affine refinement of each of the L motion vectors using an image intensity gradient-based approach.

3. Concatenation of the initial translational and the affine refinement parameters.

4. Selection of one candidate among the L estimated affine motion parameter sets.

For the first step, block matching in the long-term memory buffer is performed in order to robustly deal with large displacements yielding L translational motion vectors. In the second step, the L translational motion vectors initialize an affine estimation routine which is based on image intensity gradients. The affine motion parameters are estimated by solving an over-determined set of linear equations so as to minimize MSE. In the third step, the resulting affine motion parameter set is obtained by a weighted summation of the initial translational motion vector and the affine motion parameters. In the last step, the optimum in terms of MSE that is measured over the pixels of the cluster is chosen among the L considered candidates. In the following, the various steps are discussed in detail.

For the *first step*, the initial motion vector estimation, two approaches are discussed:

- *cluster-based initialization* and
- *macroblock-based initialization.*

For the *cluster based initialization*, the MSE for block matching is computed over all pixels inside the cluster. The motion search proceeds over the search range of ± 16 pixels and produces one motion vector per reference frame and cluster. Hence, the number of considered candidates per cluster L is equal to the number of decoded reference frames K. This approach provides flexibility in the choice of the cluster size and with that the number of clusters N. Hence, it will be used in Section 4.3 to analyze the trade-off between rate-distortion performance and complexity that is proportional to the number of initial clusters N since this number is proportional to the number of warped reference frames.

However, the *cluster-based initialization* approach produces a computational burden that increases as the number of decoded reference frames K grows since the affine refinement routine has to be repeated for each initial translational motion vector. On the other hand, translational motion estimation has to be conducted anyway for 16×16 blocks in H.263 and the long-term memory MCP coder. Hence, the re-use of those motion vectors would not only avoid an extra block matching step for the initializations, it would also fix the number of initial motion vectors to the number of macroblocks per cluster. This approach is called the *macroblock-based initialization*. Therefore, an image partitioning is considered where the clusters are aligned with the macroblock boundaries. An example for such an initial partitioning is depicted in Fig. 4.4. Fig. 4.4 shows a QCIF picture from the sequence *Foreman* that is superimposed with 99 blocks of size 16×16 pixels. The $N = 20$ clusters are either blocks of size 32×32 pixels comprising 4 macroblocks, or blocks of size 32×48, 48×32, or 48×48 pixels. If the motion vector of each macroblock is utilized as an

Figure 4.4. Image partitioning of a QCIF frame of the sequence *Foreman* into N =20 cluster.

initialization to the affine refinement step, either $L = 4, 6$ or 9 candidates have to be considered. This number is independent from the number of decoded reference frames K. The motion estimation for the macroblocks proceeds by minimizing (2.8) using the SSD distortion measure for the search range

$$\mathcal{M} = [-16 \ldots 16] \times [-16 \ldots 16] \times [1 \ldots K]. \tag{4.4}$$

followed by half-pixel refinement.

For the *second step*, the affine refinement, the initial translational motion vector $\boldsymbol{m}^I = (m_x^I, m_y^I, m_t^I)$ which is either obtained via the *cluster-based* or *macroblock-based initialization* is used to motion-compensate the past decoded frame $\acute{s}[x, y, t - m_t]$ towards the current frame $s[x, y, t]$ as follows

$$\hat{s}[x, y, t] = \acute{s}[x - m_x^I, y - m_y^I, t - m_t^I]. \tag{4.5}$$

This motion compensation has to be conducted only for the pixels inside the considered cluster $\mathcal{A}$. The minimization criterion for the affine refinement step reads as follows

$$\boldsymbol{a}^R = \underset{\boldsymbol{a}}{\operatorname{argmin}} \sum_{x,y \in \boldsymbol{\mathcal{A}}} u^2[x,y,t,\boldsymbol{a}] \tag{4.6}$$

with

$$u[x,y,t,\boldsymbol{a}] = s[x,y,t] - \hat{s}[x - m_x[\boldsymbol{a},x,y], y - m_y[\boldsymbol{a},x,y], t] \tag{4.7}$$

and $m_x[\boldsymbol{a},x,y]$ as well as $m_y[\boldsymbol{a},x,y]$ being given via (4.1).

The signal $\hat{s}[x - m_x[\boldsymbol{a},x,y], y - m_y[\boldsymbol{a},x,y], t]$ is linearized around the spatial location (x, y) for small spatial displacements $(m_x[\boldsymbol{a},x,y], m_y[\boldsymbol{a},x,y])$ yielding

$$\begin{aligned} &\hat{s}[x - m_x[\boldsymbol{a},x,y], y - m_y[\boldsymbol{a},x,y], t] \approx \\ &\hat{s}[x,y,t] - \frac{\partial \hat{s}[x,y,t]}{\partial x} m_x[\boldsymbol{a},x,y] - \frac{\partial \hat{s}[x,y,t]}{\partial y} m_y[\boldsymbol{a},x,y]. \end{aligned} \tag{4.8}$$

Hence, the error signal in (4.7) reads

$$\begin{aligned} &u[x,y,t,\boldsymbol{a}] \approx \\ &s[x,y,t] - \hat{s}[x,y,t] + \frac{\partial \hat{s}[x,y,t]}{\partial x} m_x[\boldsymbol{a},x,y] + \frac{\partial \hat{s}[x,y,t]}{\partial y} m_y[\boldsymbol{a},x,y]. \end{aligned} \tag{4.9}$$

Plugging (4.1) into (4.9) and rearranging leads to the following linear equation with 6 unknowns

$$\begin{aligned} &u[x,y,t,\boldsymbol{a}] \approx s[x,y,t] - \hat{s}[x,y,t] + \\ &(g_x c_1, g_x c_2 x', g_x c_3 y', g_y c_1, g_y c_2 x', g_y c_3 y') \begin{pmatrix} a_1 \\ a_2 \\ a_3 \\ a_4 \\ a_5 \\ a_6 \end{pmatrix} \end{aligned} \tag{4.10}$$

with the abbreviations

$$\begin{aligned} g_x &= \left(\frac{w-1}{2}\right) \frac{\partial \hat{s}[x,y,t]}{\partial x}, \qquad & g_y &= \left(\frac{h-1}{2}\right) \frac{\partial \hat{s}[x,y,t]}{\partial y}, \\ x' &= \left(x - \frac{w-1}{2}\right), & y' &= \left(y - \frac{h-1}{2}\right). \end{aligned} \tag{4.11}$$

Setting up this equation at each pixel position inside the cluster leads to an over-determined set of linear equations that is solved so as to minimize the average squared motion-compensated frame difference. In this work, the pseudo inverse technique is used which is implemented via singular value decomposition. The linearization (4.8) holds for small displacements only which might require an iterative approach to solve (4.10). However, due to the translational initialization and the subsequent quantization of the affine motion parameters it turns out that no iteration is needed. Experiments verify this statement, where the number of iterations have been varied without observing a significant difference in resulting rate-distortion performance.

The spatial intensity gradients are computed following [HS81, Hor86]. With $z \in \{x, y\}$ the spatial gradients are given as

$$\frac{\partial \hat{s}[x,y,t]}{\partial z} = \frac{1}{4}\sum_{i=0}^{1}\sum_{j=0}^{1} a_{ij}^{z} s[x+i, y+j, t] + b_{ij}^{z} \hat{s}[x+i, y+j, t], \tag{4.12}$$

with a_{ij}^z as well as b_{ij}^z being the element on the ith row and jth column of the matrices

$$\boldsymbol{A}^x = \boldsymbol{B}^x = \begin{pmatrix} 1 & -1 \\ 1 & -1 \end{pmatrix} \quad \text{and} \quad \boldsymbol{A}^y = \boldsymbol{B}^y = \begin{pmatrix} 1 & 1 \\ -1 & -1 \end{pmatrix} \tag{4.13}$$

The estimates provide the gradient of the point in-between the four samples and between the pre-compensated and the current image [Hor86]. Since the spatial gradients are computed between the pixel positions, the frame difference $s[x,y,t] - \hat{s}[x,y,t]$ is computed as well using the summation on the right hand side of (4.11) with $z = t$ and

$$\boldsymbol{A}^t = -\boldsymbol{B}^t = \begin{pmatrix} 1 & 1 \\ 1 & 1 \end{pmatrix}. \tag{4.14}$$

In the *third step*, the affine motion parameters for motion compensation between the reference frame $\acute{s}[x, y, t - m_t^I]$ and the current frame $s[x, y, t]$ is obtained via concatenating the initial translational motion vector $\boldsymbol{m}^I$ and the estimated affine motion parameter set $\boldsymbol{a}^R$ yielding

$$\begin{aligned} a_1 &= \frac{2m_x^I}{c_1(w-1)} + a_1^R, \qquad a_2 = a_2^R, \qquad a_3 = a_3^R \\ a_4 &= \frac{2m_y^I}{c_1(h-1)} + a_4^R, \qquad a_5 = a_5^R, \qquad a_6 = a_6^R \end{aligned} \tag{4.15}$$

The initial translational block matching and the affine refinement procedure are repeated for each of the L candidates. Finally, in the *fourth step*, the affine motion parameter set is chosen that minimizes the MSE measured over the pixels in the cluster $\mathcal{A}$.

4.2.2 REFERENCE PICTURE WARPING

For each of the N estimated affine motion parameter sets, the corresponding reference frame is warped towards the current frame. The reference picture warping is conducted using the motion field that is computed via (4.1) given each affine motion parameter set for the complete frame. Intensity values that correspond to non-integer displacements are computed using cubic spline interpolation [Uns99] which turns out to be more efficient than bi-linear interpolation as the motion model becomes more sophisticated [DS95]. Hence, the multi-frame buffer is extended by N new reference frames that can be used for block-based prediction of the current frame as illustrated in Fig. 4.1.

4.2.3 RATE-CONSTRAINED MULTI-FRAME HYBRID VIDEO ENCODING

At this point it is important to note that the multi-frame buffer is filled with the K most recent frames and N warped frames yielding a total of M reference frames. In order to produce the MCP signal, multi-frame block-based motion compensation is conducted. That is, half-pixel accurate motion vectors $\boldsymbol{m} = (m_x, m_y, m_t)^T$ are applied to compensate blocks of size 16×16 pixels referencing one of the $M = K + N$ reference frames. Again, block-based motion estimation is conducted to obtain the motion vectors by minimizing (2.8) as it was done when searching decoded frames to initialize affine motion estimation. In case the *macroblock-based initialization* is employed, the corresponding motion vectors can be re-used. Otherwise, motion estimation over the K decoded frames has to be conducted as described for the *macroblock-based initialization*. When searching a warped reference frame, only a range of $[-2 \ldots 2] \times [-2 \ldots 2]$ spatially displaced pixels is considered. This small search range is justified by the fact that the warped frames are already motion-compensated and experiments with a larger search range show that only a very small percentage of motion vectors is found outside the $[-2 \ldots 2] \times [-2 \ldots 2]$ range.

Given the motion vectors, the Lagrangian costs for the macroblock modes INTER, SKIP, and INTRA are computed similar to the TMN-10 specifications and the best coding options are chosen for each macroblock given the set of M reference frames. During the minimization, the values that correspond to the best coding option for a given reference frames are stored in an array. This is done to permit a fast access to the Lagrangian costs for the following step, where the number of efficient reference frames is determined.

4.2.4 DETERMINATION OF THE NUMBER OF EFFICIENT REFERENCE FRAMES

As mentioned before, there is still an open problem about the efficient combination of motion vectors, macroblock modes and reference frames. Because of the inter-dependency of the various parameters, a locally optimal solution is searched using the pre-computed Lagrangian costs. The greedy optimization algorithm proceeds as follows:

1. Sort the $M = K + N$ reference frames according to the frequency of their selection.

2. Starting with the least popular frame, test the efficiency of each reference frame by

   (a) Computing its best replacement among the more popular frames in terms of rate-distortion costs block by block.

   (b) If the costs for transmitting the reference frame parameters exceed the cost of using the replacements for this frame, remove the frame, otherwise keep it.

The first step is conducted because of the use of the variable length code to index the reference frames. The chosen reference frame with associated warping parameters are transmitted in the header of each picture. The order of their transmission provides the corresponding index that is used to specify a particular reference frame using the block-based motion vectors. This index is entropy-coded using a variable length code and the sorting matches the selection statistics to the length of the code words.

In the second step, the utility of each reference frame is tested by evaluating the rate-distortion improvement obtained by removing this reference frame. For those blocks that reference the removed frame, the best replacements in terms of Lagrangian costs among the more popular reference frames are selected. Only the more popular frames are considered because they potentially correspond to a smaller rate and because of the goal to obtain a reduced number of reference frames in the end. If no rate-distortion improvement is observed, the frame is kept in the reference buffer and the procedure is repeated for the next reference frame.

After having determined the number of efficient frames M^* in the multiple reference frame buffer, the rate-distortion costs of the INTER-4V macroblock mode are considered and the selected parameters are encoded. Up to this point, the INTER-4V mode has been intentionally left out of the encoding because of the associated complexity to determine the associated Lagrangian cost function.

4.3 EXPERIMENTS

Within the framework of the multi-frame affine motion coder there are several parameters that can be adjusted. In this section, empirical justifications are given for choices made for important parameters. Attention is given to parameters that have the largest impact on the trade-off between rate-distortion performance and computational complexity. Regarding the affine motion coder, the important question about the number of initial clusters N is discussed. This parameter is very critical since the number of warped reference pictures is directly affected by N. Then, the combination of long-term memory prediction with affine motion compensation is presented and the gains when combining affine and long-term memory MCP are investigated.

4.3.1 AFFINE MOTION COMPENSATION

In this section, the parameter setting for the affine motion coder is investigated. For that, the warping is restricted to exclusively reference the prior decoded picture. As shown later, the results for this case also propagate to a setting where the affine motion coder is combined with long-term memory prediction.

The first question to clarify concerns the number of initial clusters N. The translational motion vector estimation is conducted using the *cluster-based initialization* as described in Section 4.2.1. The coder is initialized with $N = 1, 2, 4, 8, 16, 32, 64,$ and 99 clusters. The partition into the N initial clusters is conducted so as to obtain equal size blocks and each of the blocks being as close as possible to a square. The translational motion vectors serve as an initialization to the affine refinement step described in Section 4.2.1. The estimated affine motion parameter sets are used to warp the previous decoded frame N times as explained in Section 4.2.2. Block-based multi-frame motion estimation and determination of the number of efficient affine motion parameter sets is conducted as described in Sections 4.2.3 and 4.2.4.

The left-hand side of Fig. 4.5 shows the average bit-rate savings for the set of test sequences in Tab. A.1. The procedure to obtain the average bit-rate savings is similar to the one utilized for the results in Fig. 3.13. The average bit-rate savings are very similar for the three different levels of reproduction quality. The number of initial clusters has a significant impact on resulting rate-distortion performance. The increase in bit-rate savings tends to a saturation for a large number of clusters, i.e., more than 32 clusters, reaching the value of 17 % for the set of test sequences considering the reproduction quality of 34 dB PSNR.

This can be explained when investigating the average number of affine motion parameter sets that are transmitted as shown on right-hand side in Fig. 4.5. The average number of transmitted affine motion parameter sets is generated with a similar method as the average bit-rate savings for a given PSNR value.

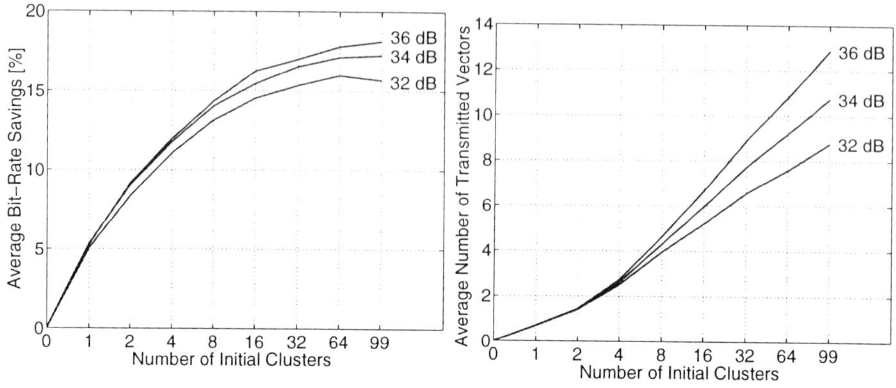

Figure 4.5. Average bit-rate savings (left) and average number of transmitted affine motion parameter sets (right) vs. number of initial clusters for the test sequences in Tab. A.1 and three different levels of reproduction quality.

The average number of affine motion parameter sets increases with increasing average PSNR as well as an increased number of initial clusters. This is because the size of the measurement window becomes smaller as the number of initial clusters increases and the affine motion parameters are more accurate inside the measurement window. Hence, the coder chooses to transmit more affine motion parameter sets. For very small numbers of initial clusters, a large percentage of the maximum number of affine motion parameter sets is chosen. However, as the number of initial clusters is increased, a decreasing percentage of affine motion parameter sets is transmitted.

Figure 4.6 shows the average bit-rate savings at 34 dB PSNR for the set of test sequences where the result for each sequence is shown. The abbreviations *fm, mc, st, te, cs, md, nw,* and *si* correspond to those in Tab. A.1. The solid line depicts the average bit-rate savings for the 8 test sequences at equal PSNR of 34 dB. The results differ quite significantly among the sequences in the test set. On the one hand, for the sequence *Silent Voice*, only a bit-rate saving of 6 % can be obtained. On the other hand, sequences like *Mobile & Calendar* and *Container Ship* show substantial gains of more than 25 % in bit-rate savings.

In Fig. 4.6, the asterisk shows the average result for the *macroblock-based initialization* of the affine estimation (see Section 4.2.1). Please recall that all experiments that were described so far are conducted using the *cluster-based initialization* for the translational motion vector estimation to have a simple means for varying the number of initial clusters. For the *macroblock-based initialization*, the segmentation in Fig. 4.4 is employed resulting in $N = 20$ clusters. The bit-rate saving of 15 % is very close to the results for the *cluster-based initialization*. However, the complexity is drastically reduced.

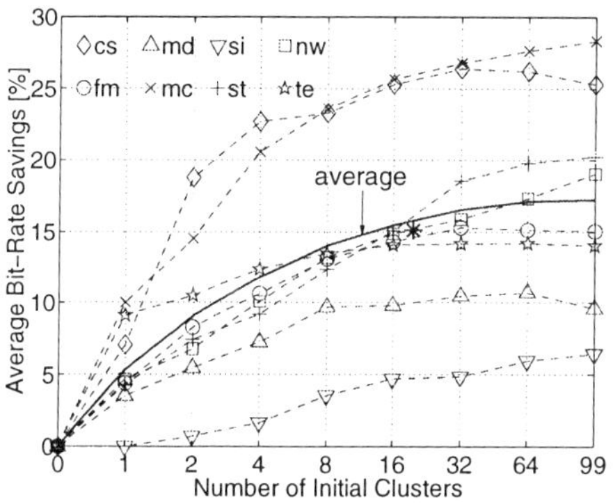

Figure 4.6. Average bit-rate savings at 34 dB PSNR versus number of initial clusters for the test sequences in Tab. A.1.

Typical running time measurements for the *macroblock-based initialization* are as follows. The complete affine motion coder runs at 6.5 seconds per QCIF frame on a 300 MHz Pentium PC. These 6.5 seconds are split into 0.5 seconds for translational motion estimation for 16×16 macroblocks, 1 second for affine motion estimation, and the warping also takes 1 second. The pre-computation of the costs for the INTER, SKIP, and INTRA mode takes 2 seconds, and the remaining steps use 2 seconds. As a comparison, the TMN-10 coder which has a similar degree of run-time optimization uses 2 seconds per QCIF frame.

Finally, rate-distortion curves are depicted to evaluate the performance of this approach. For that, the DCT quantization parameter has been varied over values $Q = 4, 5, 7, 10, 15,$ and 25 when encoding the sequences *Foreman*, *Mobile & Calendar*, *News*, and *Tempete*. The results are shown in Fig. 4.7, where the rate-distortion curves for the affine motion coder are compared to those of TMN-10 when running both codecs according to the conditions in Tab. A.1. The following abbreviations indicate the two codecs compared:

- **TMN-10:** The H.263 test model using Annexes D, F, I, J, and T.
- **MRPW:** As TMN-10, but motion compensation is extended to referencing warped frames corresponding to $N = 20$ initial clusters using the *macroblock-based initialization*.

The PSNR gains vary for the different test sequences and tend to be larger as the bit-rate increases. In contrast, the relative bit-rate savings are more or less constant over the entire range of bit-rates that was tested. Typically, a PSNR gain of 1 dB compared to TMN-10 is obtained. The PSNR gains are up to 2.3 dB for the sequence *Mobile & Calendar*.

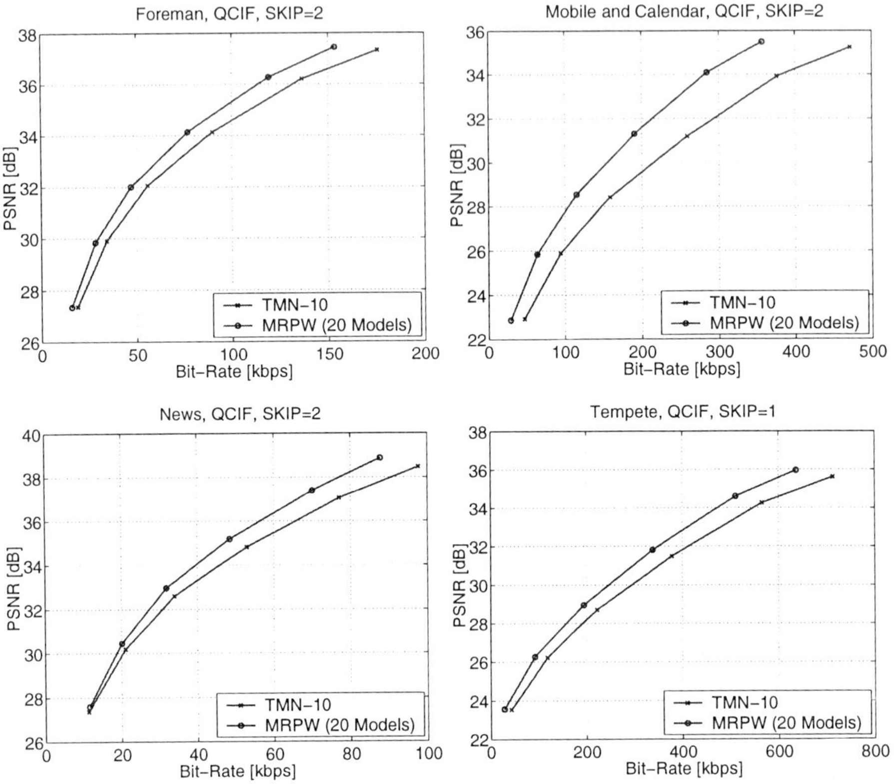

Figure 4.7. PSNR vs. overall bit-rate for the QCIF sequences *Foreman* (top left), *Mobile & Calendar* (top right), *News* (bottom left), and *Tempete* (bottom right).

4.3.2 COMBINATION OF AFFINE AND LONG-TERM MEMORY MOTION COMPENSATION

In the previous section, it is shown that affine motion compensation provides significant bit-rate savings. The gains for the affine motion coder increase with an increasing number of initial clusters. A saturation of the gains is reported when increasing the number of initial clusters beyond 32. The number of initial clusters determines the number of reference frames that are warped. Hence, a parameter choice is proposed where 20 initial clusters are utilized providing an average bit-rate saving of 15 %.

In contrast to the affine motion coder where warped versions of the prior decoded frame are employed, the long-term memory MCP coder references past decoded frames for motion compensation. However, aside from the different origin of the various reference frames, the syntax for both codecs is very similar. In Chapter 3, Fig. 3.13 shows the average bit-rate savings at 34 dB PSNR for the set of test sequences that are achieved with the long-term memory MCP codec.

In Chapter 3, it is found that long-term memory MCP with 10 past decoded frames for most sequences yields a good compromise between complexity and bit-rate savings.

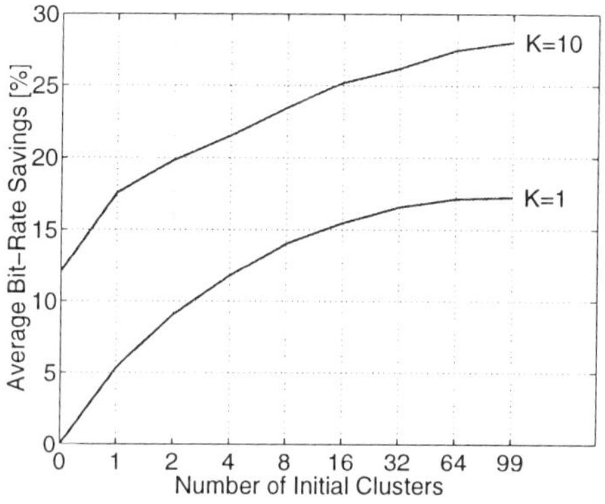

Figure 4.8. Average bit-rate savings at 34 dB PSNR versus number of initial clusters for the set of test sequences in Tab. A.1. Two cases are shown: (*i*) affine warping using $K = 1$ reference frames (lower solid curve) and (*ii*) affine warping using $K = 10$ reference frames (upper solid curve).

In Fig. 4.8, the result is depicted when combining the affine motion coder and long-term memory MCP. This plot shows average bit-rate savings at 34 dB PSNR versus the number of initial clusters for the set of test sequences in Tab. A.1. Two cases are shown: (*i*) affine warping using $K = 1$ reference frame (lower solid curve) and (*ii*) affine warping using $K = 10$ reference frames (upper solid curve). For the case $K = 1$, the setting of the coder has been employed again that was used for the curve depicting the average bit-rate savings at 34 dB on the left-hand side in Fig. 4.5. To obtain the result for the case $K = 10$, the combined coder is run using the *cluster-based initialization* with $N = 1, 2, 4, 8, 16, 32, 64$, and 99 initial clusters. For the *cluster-based initialization* of the affine motion estimation, $L = K = 10$ initial translational motion vectors are utilized each corresponding to the best match on one of the K decoded frames (see Section 4.2.1). Please note that the number of maximally used reference frames is $N + K$. Interestingly, the average bit-rate savings obtained by the affine motion and the long-term memory prediction coder are almost additive when being combined using multi-frame affine MCP.

Figure 4.9 shows the bit-rate savings for each of the test sequences in Tab. A.1 when employing $K = 10$ decoded reference frames versus the number of initial clusters N using dashed lines. The solid line for $K = 10$ is repeated from the left-hand side of Fig. 4.5. The bit-rate savings are more than 35 % for the sequences *Container Ship* and *Mobile & Calendar* when using 32 or more initial clusters. Interestingly, when using $K = 10$ reference frames and 16 or more initial clusters the bit-rate savings are never below 17 %.

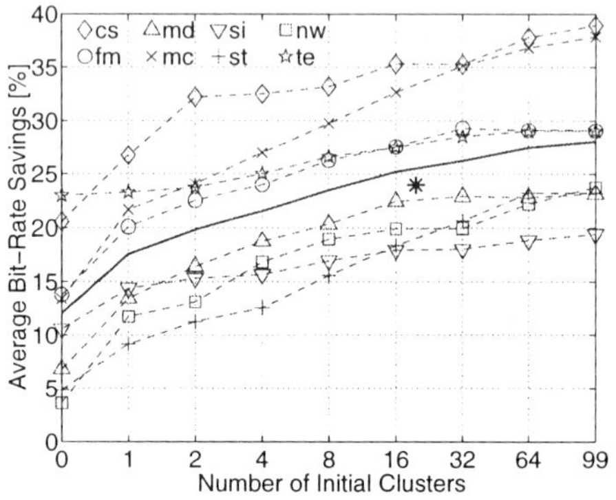

Figure 4.9. Average bit-rate savings at 34 dB PSNR versus number of initial clusters for the set of test sequences in Tab. A.1.

In Fig. 4.9, the asterisk shows the result for the case of *macroblock-based initialization*. For that, the initial segmentation in Fig. 4.4 is employed. The initial motion vectors for the affine motion estimation are those best matches found for the macroblocks in each cluster when searching $K = 10$ decoded reference frames. An average bit-rate saving of 24 % is obtained for the set of 8 test sequences in Tab. A.1.

The measured bit-rate savings correspond to PSNR gains up to 3 dB. Figure 4.10 shows rate-distortion curves for the four test sequences *Foreman*, *Mobile & Calendar*, *Container Ship*, and *Silent Voice*. The curves depict the results that are obtained with the following three codecs:

- **TMN-10:** The H.263 test model using Annexes D, F, I, J, and T.
- **LTMP:** As TMN-10, but motion compensation is extended to long-term memory prediction with $K = 10$ decoded reference frames.
- **MRPW+LTMP:** As TMN-10, but motion compensation is extended to combined affine and long-term memory prediction. The size of the long-term memory is selected as $K = 10$ frames. The number of initial clusters is $N = 20$ and the *macroblock-based initialization* is employed.

Long-term memory MCP with $K = 10$ frames and without affine warping is always better than TMN-10 as already demonstrated in Chapter 3. Moreover, long-term memory MCP in combination with affine warping is always better than the case without affine warping. Typically, bit-rate savings between 20 and 35 % can be obtained which correspond to PSNR gains of 2-3 dB. For

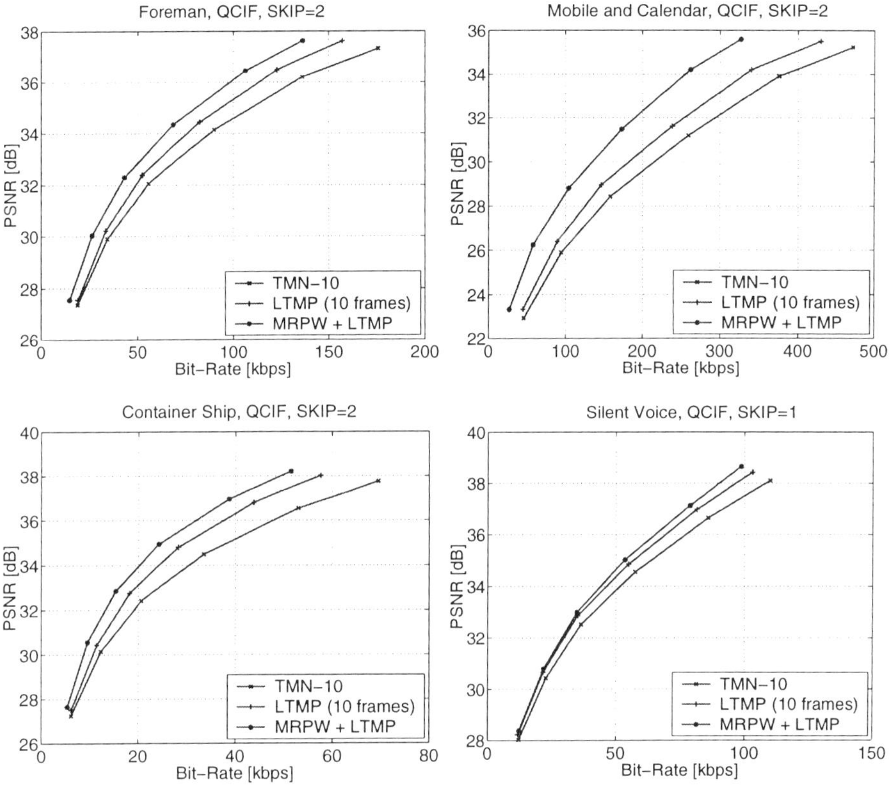

Figure 4.10. PSNR vs. overall bit-rate for the QCIF sequences *Foreman* (top left), *Mobile & Calendar* (top right), *Container Ship* (bottom left), and *Silent Voice* (bottom right).

some sequences long-term memory prediction provides most of the gain (*Silent Voice*) while for other sequences the affine motion coder is more important (*Mobile & Calendar*).

For the sequence *Mobile & Calendar* the gap between the result for the long-term memory MCP codec with and without affine motion compensation is visible for the lowest bit-rates as well. This results in a bit-rate saving of 50 %. Moreover, for some sequences, the gain obtained by the combined coder is larger than the added gains of the two separate coders. For example, the long-term memory prediction gain for *Mother & Daughter* is 7 % for $K = 10$ reference pictures when measuring over all coded frames. The gain obtained for the affine motion coder is 10 % when using 32 initial clusters. However, the combined coder achieves 23 % bit-rate savings for the sequence *Mother & Daughter*.

4.4 ASSESSMENT OF THE RATE-DISTORTION PERFORMANCE OF MULTI-FRAME PREDICTION

In this section, the bit-rate reduction and the PSNR gain of multi-frame MCP is compared against what has been achieved using improved motion compensation in the past. In Chapter 1, Fig. 1.4 depicts the average bit-rate savings versus increased prediction capability for the test sequences in Tab. A.1. Figure 4.11 shows the rate-distortion curves for the sequences *Foreman*, *Mobile & Calendar*, *Container Ship*, *Silent Voice*. The experimental conditions are those from Tab. A.1 and the acronyms CR, FD, IP-MC, HP-MC, TMN-10 correspond to those in Fig. 1.4. The result for multi-frame prediction is labeled by MFP and corresponds to those performance measures as in Fig. 4.10, where $K = 10$ reference frames and $N = 20$ initial clusters are utilized.

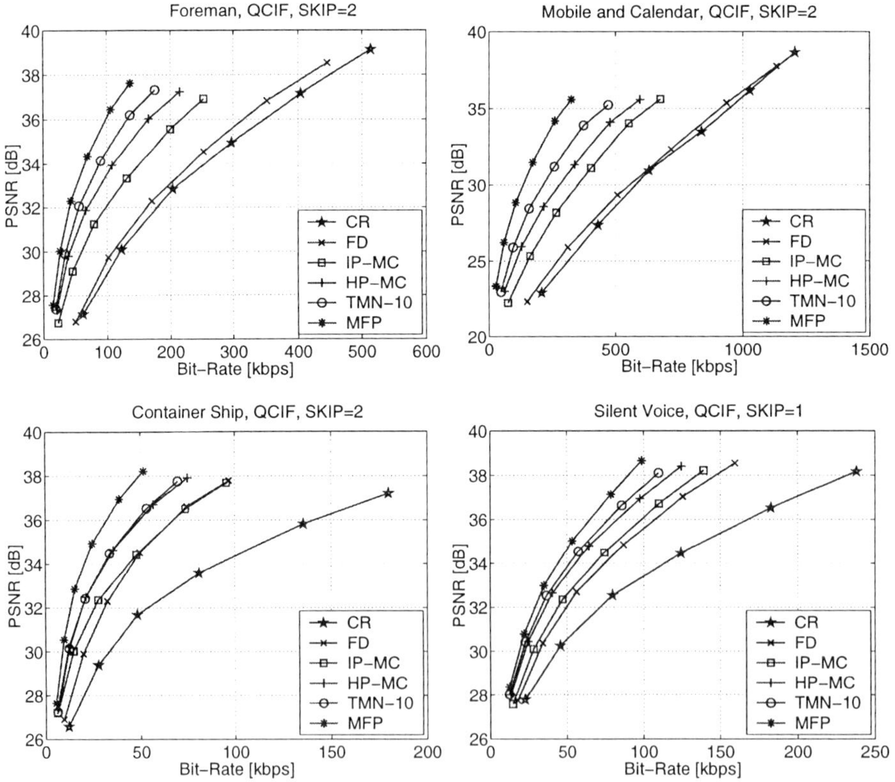

Figure 4.11. PSNR vs. overall bit-rate for the QCIF sequences *Foreman* (top left), *Mobile & Calendar* (top right), *Container Ship* (bottom left), and *Silent Voice* (bottom right).

For all sequences, the combined affine and long-term memory MCP codec makes a visible difference and the result for MFP can be easily distinguished from the state-of-the-art. The bit-rate savings obtained by multi-frame prediction against the TMN-10 coder in Fig. 4.9 is 24 % when setting the bit-rate of the TMN-10 case to 100 %. Relative to the bit-rate for integer-pixel motion compensation, half-pixel motion provides 21 % bit-rate savings which corresponds to the gain when using H.263-baseline prediction instead of H.261-based motion compensation (without a loop filter [ITU93]). Hence, the bit-rate savings obtained with multi-frame affine motion compensation are larger than those obtained when improving the MCP from H.261 to H.263 in baseline mode.

4.5 DISCUSSION AND OUTLOOK

The affine motion model can be viewed as an extended set of basis functions for the representation of the motion vector field in the image plane when comparing it to the translational motion model. Hence, improvements beyond the results with the affine motion model could be obtained when further extending the set of basis functions. The chosen set of basis functions must be suitable for an efficient estimation approach of the motion coefficients and the image segmentation. When more basis functions are considered, advanced techniques for the coding of the motion coefficients become increasingly important.

Another multi-frame MCP scheme is presented in [EWG00], where a hybrid video coder is combined with a 3-D model-based approach for efficient compression of head-and-shoulder sequences. The combination with multi-frame prediction is achieved by running a block-based two-frame video coder with the prior decoded picture and a synthesized picture by the model-based coder. The model-based coder uses a parameterized 3-D head model specifying shape and color of a person. The transmitted parameters of the 3-D head model are estimated at the encoder using the current picture and the prior picture. Both approaches are combined using the rate-constrained multi-frame MCP approach as presented in the previous chapter. Hence, the generality of waveform coding and the efficiency of 3-D model-based coding are available where needed. Experiments on five video sequences show that bit-rate savings of about 35 % are achieved at equal average PSNR when comparing the coder in [EWG00] to TMN-10, the state-of-the-art test model of the H.263 standard. This corresponds to a gain of 2-3 dB in PSNR when encoding at the same average bit-rate [EWG00].

4.6 CHAPTER SUMMARY

The idea of reference picture warping can be regarded as an alternative approach to assigning affine motion parameters to large image segments with the aim of a rate-distortion efficient motion representation. Although the affine motion parameter sets are determined on sub-areas of the image, they can be employed at any position inside the frame. Instead of performing a joint estimation of the image partition and the associated affine motion parameter sets, reference frames are warped and selected in a rate-distortion efficient way on a block basis. Hence, the presented approach decomposes the joint optimization task of finding an efficient combination of affine motion parameters, regions and other parameters into separate steps. Each of these steps takes an almost constant amount of computation time which is independent of the context of the input data. The coder robustly adapts the number of affine motion parameter sets to the input statistics and never degrades below the rate-distortion performance that can be achieved with the syntax of the underlying H.263 standard. The use of multiple reference frames requires only very minor syntax changes of state-of-the-art video coding standards.

The combined affine and long-term memory MCP codec is an example for an efficient multi-frame video compression scheme. The two incorporated multi-frame concepts seem to complement each other well providing almost additive rate-distortion gains. When warping the prior decoded frame, average bit-rate savings of 15 % against TMN-10 are reported for the case that 20 warped reference pictures are used. For the measurements, reconstruction PSNR is identical to 34 dB for all cases considered. These average bit-rate savings are measured over a set of 8 test sequences that represent a large variety of video content. Within the test set, the bit-rate savings vary from 6 to 25 %. Long-term memory prediction has been already demonstrated as an efficient means to compress motion video. The efficiency in terms of rate-distortion performance is comparable to that of the affine coder. The combination of the two approaches yields almost additive average gains. When employing 20 warped reference pictures and 10 decoded reference frames, average bit-rate savings of 24 % can be obtained for the set of 8 test sequences. The minimal bit-rate savings inside the test set are 15 % while the maximal bit-rate savings are reported to be up to 35 %. These bit-rate savings correspond to gains in PSNR between 0.8 and 3 dB. For some cases, the combination of affine and long-term memory MCP provides more than additive gains.

Chapter 5

FAST MOTION ESTIMATION FOR MULTI-FRAME PREDICTION

Multi-frame MCP can provide substantial improvements in rate-distortion performance. These improvements come with the following drawbacks

1. increased memory requirement at the encoder and decoder,

2. increased computational complexity at the encoder.

The first issue is not considered here since memory is increasingly a commodity, a fact which motivated the investigations in this book. The second item, the increased computational complexity at the encoder due to multi-frame motion estimation is still important in today's video encoders. This chapter focuses on methods for lowering the computation time of the motion estimation.

Multi-frame block-based motion estimation is conducted by block matching in the search space which is for QCIF pictures typically set to ± 16 pixels horizontally and vertically for each reference picture. The main idea investigated in this chapter is to pre-compute data about the search space that can be used to either avoid considering certain positions or to reduce the complexity for evaluating distortion. For that, it is important to consider the computation time for the pre-computation step to arrive at a lower complexity overall.

The multi-frame scenario that is considered here is long-term memory MCP employing a sufficiently large number of reference frames ($M \geq 10$). When a picture is decoded and included into the set of M reference pictures, it is assumed that most of the pictures remain unchanged in the buffer which is the case for the *Sliding Window* buffering rule. This condition distinguishes the fast search methods for multi-frame MCP from most single-frame approaches because the relative portion of time for the pre-computation step becomes smaller as the number of unchanged reference frames increases.

This chapter is organized as follows. In Section 5.1, strategies for the reduction of computation time for multi-frame block matching are presented that do not introduce any loss of rate-distortion performance in comparison to the full search approach in Chapter 3. Further reductions in computation time can be obtained by accepting some losses in rate-distortion performance. Section 5.2 presents such ideas. Finally, in Section 5.3, experimental results illustrate the particular benefits for the various search strategies.

5.1 LOSSLESS FAST MOTION ESTIMATION

For each position in the search space, the costs that determine the criterion for motion estimation have to be computed and the optimum among those candidates is chosen. In this work, rate-constrained motion estimation is utilized, where the criterion to find the optimum motion vector $\boldsymbol{m}_i$ for the block $\boldsymbol{S}_i$ is the minimization of a Lagrangian cost function

$$J_{\text{MOTION}}(\boldsymbol{\mathcal{S}}_i, \boldsymbol{m}) = D_{\text{DFD}}(\boldsymbol{\mathcal{S}}_i, \boldsymbol{m}) + \lambda_{\text{MOTION}} R_{\text{MOTION}}(\boldsymbol{\mathcal{S}}_i, \boldsymbol{m}), \qquad (5.1)$$

for each candidate $\boldsymbol{m}$ in the search space. The distortion term $D_{\text{DFD}}(\boldsymbol{\mathcal{S}}_i, \boldsymbol{m})$ is being measured as SAD or SSD, and $R_{\text{MOTION}}(\boldsymbol{\mathcal{S}}_i, \boldsymbol{m})$ is the rate associated with the motion vector $\boldsymbol{m}$. Typically, the rate for the motion vector is computed by a table look-up which is comparably fast. However, the computation of the distortion for the various search positions is demanding. Hence, fast search methods for motion estimation attempt the reduction of computation time for the distortion term in (5.1).

The main principle that is employed in this chapter to reduce computation time is based on excluding candidates $\boldsymbol{m}$ or terminating the distortion computation for them early. Let us consider J_{min} as the minimum Lagrangian cost value that has been determined so far in the motion search. Any candidate $\boldsymbol{m}$ can not provide a lower value than J_{min} if a Lagrangian cost function that incorporates an approximate distortion measure $D'_{\text{DFD}}(\boldsymbol{\mathcal{S}}_i, \boldsymbol{m})$ that satisfies the following condition

$$D'_{\text{DFD}}(\boldsymbol{\mathcal{S}}_i, \boldsymbol{m}) \leq D_{\text{DFD}}(\boldsymbol{\mathcal{S}}_i, \boldsymbol{m}), \qquad (5.2)$$

exceeds J_{min}, i.e.

$$D'_{\text{DFD}}(\boldsymbol{\mathcal{S}}_i, \boldsymbol{m}) + \lambda_{\text{MOTION}} R_{\text{MOTION}}(\boldsymbol{\mathcal{S}}_i, \boldsymbol{m}) \geq J_{\text{min}}. \qquad (5.3)$$

This method will never provide inferior results in terms of Lagrangian costs, if the exclusion of candidates is based on approximations that provide smaller values than the actual Lagrangian cost value of that candidate. Such smaller values are available during SAD or SSD computation, since the distortion per pixel is a positive quantity and the summation of positive values is a monotonically increasing function. Typically, these partial distortion measures are computed after each row of 16×16 or 8×8 blocks. Thus, this method serves as the anchor in all comparisons.

Other approximations of the distortion measure that are based on triangle inequalities are described in Section 5.1.1. The other side of the inequality in (5.3) is important as well, because if $J_{\min}$ is small in the beginning of the search, more candidates can be rejected. This effect can be controlled by the search order which is described in Section 5.1.2. Another important aspect is the number of blocks that are considered. Hence, the search space for motion estimation is discussed in Section 5.1.3.

5.1.1 TRIANGLE INEQUALITIES FOR DISTORTION APPROXIMATION

An approximation of the distortion term in (5.3) that satisfies (5.2) is given by the triangle inequality, which provides a lower bound on the norm of the difference between vectors [LS95, LT97]. The special structure of the motion estimation problem permits a fast method to pre-compute the norm values of all blocks in the previously decoded frames [LS95]. Incorporating the triangle inequality into the sums for SAD and SSD, yields

$$
\begin{aligned}
D_{\mathrm{DFD}}(\boldsymbol{S}_i, \boldsymbol{m}) &= \sum_{(x,y)\in\boldsymbol{\mathcal{A}}_i} \Big| s[x,y,t] - \acute{s}[x-m_x, y-m_y, t-m_t] \Big|^p \geq \\
D'_{\mathrm{DFD}}(\boldsymbol{S}_i, \boldsymbol{m}) &= \qquad\qquad\qquad\qquad (5.4)\\
&\left| \left(\sum_{(x,y)\in\boldsymbol{\mathcal{A}}_i} \Big| s[x,y,t] \Big|^p \right)^{1/p} - \left(\sum_{(x,y)\in\boldsymbol{\mathcal{A}}_i} \Big| \acute{s}[x-m_x, y-m_y, t-m_t] \Big|^p \right)^{1/p} \right|^p
\end{aligned}
$$

by varying the parameter $p = 1$ for SAD and $p = 2$ for SSD. The set $\boldsymbol{\mathcal{A}}_i$ comprises the sampling positions of the blocks considered, e.g., a block of 16×16 samples.

The concept of one triangle inequality per block can be extended to multiple triangle inequalities. Assume a partition of the set $\boldsymbol{\mathcal{A}}_i$ into subsets $\boldsymbol{\mathcal{A}}_i^n$ such that

$$
\boldsymbol{\mathcal{A}}_i = \bigcup_n \boldsymbol{\mathcal{A}}_i^n, \qquad \text{and} \qquad \bigcap_n \boldsymbol{\mathcal{A}}_i^n = \emptyset. \tag{5.5}
$$

The triangle inequality (5.3) holds for all possible subsets $\boldsymbol{\mathcal{A}}_i^n$. Thus, rewriting the formula for $D_{\mathrm{DFD}}(\boldsymbol{S}_i, \boldsymbol{m})$ yields

$$
\begin{aligned}
&\sum_{(x,y)\in\boldsymbol{\mathcal{A}}_i} \Big| s[x,y,t] - \acute{s}[x-m_x, y-m_y, t-m_t] \Big|^p = \\
&\sum_n \sum_{(x,y)\in\boldsymbol{\mathcal{A}}_i^n} \Big| s[x,y,t] - \acute{s}[x-m_x, y-m_y, t-m_t] \Big|^p \qquad (5.6)
\end{aligned}
$$

and applying the triangle inequality for all $\boldsymbol{\mathcal{A}}_i^n$ provides a lower bound for the distortion as follows

$$D_{\mathrm{DFD}}(\boldsymbol{\mathcal{S}}_i, \boldsymbol{m}) \geq \tag{5.7}$$
$$\sum_n \left| \left(\sum_{(x,y) \in \boldsymbol{\mathcal{A}}_i^n} \left| s[x,y,t] \right|^p \right)^{1/p} - \left(\sum_{(x,y) \in \boldsymbol{\mathcal{A}}_i^n} \left| \acute{s}[x - m_x, y - m_y, t - m_t] \right|^p \right)^{1/p} \right|^p$$

Note that (5.7) is a tighter lower bound than (5.3), but it requires more computation. Hence, a trade-off can be obtained between the sharpness of the lower bound and computational complexity.

An important issue within this context remains to be the choice of the partitions $\boldsymbol{\mathcal{A}}_i^n$. Of course, (5.7) works for all possible subsets that satisfy (5.5). However, since the norm values of all blocks in the search space have to be pre-computed, the fast method as described in [LS95] should be used. Therefore, a random sub-division of $\boldsymbol{\mathcal{A}}_i$ into n subsets may not be the appropriate choice. Instead, for sake of computation, a regular sub-division of $\boldsymbol{\mathcal{A}}_i$ is more desirable.

Please recall that the H.263 video coding standard permits blocks of size 16×16 and blocks of size 8×8 if the Annexes F or J are enabled as is the case in our experiments. Hence, the idea proposed in [LC95] is employed where a 16×16 block is decomposed into 4 different partitions. The 16×16 block is partitioned into 1 set of 16×16 samples, into 4 subsets of 8×8 samples, into 16 subsets of 4×4 samples, and into 64 subsets of size 2×2 samples. The various triangle inequalities are successively applied in the order of the computation time to evaluate them, i.e., first the 16×16 triangle inequality (5.3) is tested, then the inequalities relating to blocks of size 8×8, 4×4, and 2×2 samples are computed using (5.7). This hierarchy of triangle inequalities is also applied for 8×8 blocks where the triangle inequality in (5.3) is first employed for the complete 8×8 block and the subset triangle inequalities in (5.7) are evaluated for 4×4 and 2×2 blocks accordingly.

5.1.2 SEARCH ORDER

It is obvious that a small value for $J_{\min}$ determined in the beginning of the search leads to the rejection of many other blocks later and thus reduces computation time. Hence, the order in which the blocks in the search space are tested has an impact on the computation time. Here, two strategies are considered:

1. probability-ordered search and
2. norm-ordered search.

The probability-ordered search follows an increasing number of bits for the motion vectors, i.e., motion vectors corresponding to a small number of bits are considered first. The idea is that the variable length code for the motion vectors corresponds to an entropy-code and with that a small number of bits is associated to a high probability that this motion vector is being transmitted. If candidates that correspond to small values for $J_{\min}$ are considered first, many candidates that are successively tested using (5.3), can be rejected.

However, the motion vector that is actually chosen for a particular block may significantly differ from those statistics that are used to derive the variable length codes. Another measure about the probabilities of finding a good match in the search space is given by the values of the block norms that are used for the triangle inequality in (5.3). This is the idea behind the norm-ordered search, where blocks in the search space having a similar norm as the block which is to be compensated are tested first [WLG98]. The norm-ordered search stops, if (5.3) is violated. Thus, the algorithm does not even have to look at those positions, which cannot yield Lagrangian costs lower than the one previously found.

5.1.3 SEARCH SPACE

The search space should cover a sufficient range of possible shifts in the image plane. Sufficient in this case means, that a further increase of the search space does not provide significant improvements in rate-distortion performance. For pictures in QCIF resolution, empirical results for the set of test sequences in Tab. A.1 show that a range of ± 16 pixels in horizontal and vertical direction is sufficient. This also holds for most sequences in the case of long-term memory MCP, when the range of ± 16 pixels is searched for each of the M reference frames. But this extended search space might introduce redundancy, i.e., the blocks in the various pictures might be similar. Moreover, in the hybrid coder, quantized pictures are used as reference for the current picture. This leads to identical pixels for blocks that are coded by the SKIP mode which is quite often the case for frames that are generated by a still camera. Typically, for sequences like *Container Ship*, *Mother & Daughter*, *News*, and *Silent Voice*, more than 50 % of a picture are coded using SKIP.

Identical blocks in the search space lead to identical distortion values. Hence, if the number of bits to reference two or more identical blocks are different, all search positions, except for the one corresponding to the minimum number of bits, can be excluded from the search space. A simple method to exclude such positions is to compute the frame difference between the most recent reference frame and the current decoded frame. For all pixel-displaced 16×16 and 8×8 blocks $\mathcal{A}_i$, the SSD is computed over the difference frame. Those positions, for which the following condition is satisfied

$$\frac{1}{|\mathcal{A}_i|} \sum_{(x,y) \in \mathcal{A}_i} (\acute{s}[x,y,t] - \acute{s}[x,y,t-1])^2 \leq \Theta \tag{5.8}$$

with $\Theta = 0$ are excluded in the older of the two reference frames, without introducing any loss in prediction performance. Values of $\Theta > 0$ lead to a further reduction of the search space, but might introduce a loss in prediction performance as discussed in the next section. The quantity $|\mathcal{A}_i|$ specifies the number of pixels per block. The positions in the older reference frame are excluded, because they require more or an equal number of bits for the picture reference parameter than the more recent frame when employing the sliding window buffering. This exclusion of search positions is only done for the most recent reference frame and the current decoded frame. The motion estimation incorporates the exclusion of search positions into the probability-ordered or norm-ordered search, by generating a mask that indicates, whether a certain position should be considered or not. This mask is attached to each reference frame and utilized as long as the corresponding frame remains in the multi-frame buffer.

5.2 LOSSY FAST MOTION ESTIMATION

In many applications the computation time available is very often not sufficient in order to conduct lossless motion estimation. Hence, the computation time needs to be further reduced, which, in general, results in losses in rate-distortion performance. Two ideas have mainly been explored in the context of lossy search methods:

1. sub-sampling of the search space,
2. sub-sampling of the block for which distortion is measured.

In the following, the application of both ideas to speed-up long-term memory MCP is explained.

5.2.1 SUB-SAMPLING OF THE SEARCH SPACE

The sub-sampling of the search space is realized by the exclusion of "too similar" or "too different" blocks from the search space. Blocks that are "too similar" are removed from the search space via setting $\Theta > 0$ in (5.8). Hence, if the difference between decoded frames is too small, the corresponding positions of the older of the two frames are excluded from the search. Such a scheme is suggested by the statistical model for the minimization of long-term memory MCP in Section 3.3. The analysis in Section 3.3 yields the insight that highly correlated distortion values in the search range provide small prediction gains. It can be shown that the correlation of the distortion values of two blocks in the search space increases when the correlation of the pixel values of those blocks increases. Hence, if the left-hand side term in (5.8) is small, the pixel values of the two considered blocks are highly correlated and with that the corresponding distortion values.

Blocks that are "too different" are removed by early termination of the norm-ordered search. Please recall that the norm-ordered search first tests candidates which have a norm value that is similar to the norm value of the block for which the search is conducted. The norm value is used in the triangle inequality (5.3) to obtain a lower bound on the true distortion. This lower bound is included into the criterion for motion estimation in (5.3) to exclude certain candidates. Hence, if (5.3) is modified to

$$D'_{\text{DFD}}(\boldsymbol{S}_i, \boldsymbol{m}) \cdot \Omega + \lambda_{\text{MOTION}} R_{\text{MOTION}}(\boldsymbol{S}_i, \boldsymbol{m}) \geq J_{\min} \tag{5.9}$$

with $\Omega > 1$, the costs of certain positions might be over-estimated because $D'_{\text{DFD}}(\boldsymbol{S}_i, \boldsymbol{m}) \cdot \Omega$ is no longer guaranteed to be a lower bound of the true distortion. Doing so, a candidate might be excluded from the search, which potentially yields lower costs than $J_{\min}$. On the other hand, the modification in (5.9) may save computation time by excluding candidates for which the lower bound on the distortion in (5.3) might be too conservative. Thus, by adjusting Ω, a trade-off can be achieved between computation time and rate-distortion performance.

5.2.2 SUB-SAMPLING OF THE BLOCK

Sub-sampling of the block for which distortion is measured, reduces the computation time that is needed to test a particular block. Often this approach is realized by evaluating the distortion criterion, using a reduced number of samples. The reduction of the number of samples can be achieved by filtering and sub-sampling. The use of the triangle inequality can be interpreted as filtering and sub-sampling. Hence, this approach to fast motion estimation can be incorporated by stopping the distortion evaluation at a certain level in the hierarchy of triangle inequalities.

The level in the hierarchy is adapted by measuring the amount of activity in the block for which the motion search is conducted. More precisely, the activity is measured as

$$\Xi \geq \tag{5.10}$$

$$\frac{1}{|\boldsymbol{\mathcal{B}}_i|}\left[\sum_{(x,y)\in\boldsymbol{\mathcal{B}}_i}(s[x+1,y,t]-s[x,y,t])^2+\sum_{(x,y)\in\boldsymbol{\mathcal{B}}_i}(s[x,y+1,t]-s[x,y,t])^2\right]$$

with Ξ being a threshold and $\boldsymbol{\mathcal{B}}_i$ being the 15×15 or 7×7 block from the upper left corner of the considered 16×16 or 8×8 blocks, respectively. If the activity measure is below the threshold Ξ, the distortion is only measured on the 2×2 block level of the triangle inequality hierarchy and never evaluated on the pixel level. Please note that the consideration of the 4×4 or larger blocks for the triangle inequality level has not provided good results and is therefore not considered.

5.3 EXPERIMENTS

In the following, the results of experiments are presented that are conducted to illustrate the particular benefits for the presented approaches to computation time reduction. For that, the TMN-10 coder and the long-term memory MCP coder are employed with Annexes D, F, I, J and T being enabled. Hence, the motion estimation is conducted for 16×16 and 8×8 blocks. The motion search range in all simulations is set to $\mathcal{M} = [-16 \ldots 16] \times [-16 \ldots 16] \times [1 \ldots M]$, with $M = 1$ for the TMN-10 coder and with $M = 10$ and $M = 50$ reference frames for the long-term memory MCP coder. The distortion criterion is the SAD measure. Comparisons are made for the computation time needed to conduct integer-pixel accurate motion estimation for 16×16 and 8×8 blocks together when employing the various approaches presented before. All remaining parts of the hybrid coder require the same amount of computation for all cases considered. The experiments are conducted on a 300 MHz Pentium PC, single processor, 256 MByte RAM, and Linux operating system. No advanced instruction sets are used. Note that the results depend on the machine used and they might be different for other platforms. But the main intention here is to illustrate the impact of the various approaches.

5.3.1 RESULTS FOR LOSSLESS METHODS

Section 5.1 describes the exclusion of candidates in the search space using distortion approximations as the main approach to fast motion estimation. An experiment has been conducted to illustrate the reductions in computation time, when using the distortion approximation with the triangle inequality, varying the search order and reducing the search space for long-term memory MCP. Figure 5.1 depicts the average computation time T in seconds per frame for the integer-pixel motion search for 16×16 and 8×8 blocks together, that has been measured by excluding the first 50 of the encoded frames for the sequences *Foreman*, *Mobile & Calendar*, *Container Ship*, and *Silent Voice*. Each sequence has been encoded according to the simulation conditions in Tab. A.1 using the following approaches:

- **FS**: Full search, i.e., no triangle inequalities are utilized. The blocks are tested using the probability-ordered search as described in Section 5.1.2. For each candidate, (5.3) is tested after each line of samples when computing the SAD of the 16×16 or 8×8 blocks.

- **POS**: Probability-ordered search. This approach is similar to FS, except that for the lower bound of the distortion D'_{DFD} in (5.3), the triangle inequality is evaluated. The time for the pre-computation of the norm values is included in the results. (This is also the case for other techniques which require pre-computation.)

- **NOS**: Norm-ordered search. As FS, but the norm-ordered search as described in Section 5.1.2 is employed. For each search candidate, (5.3) is tested in order to exclude candidates based on their rate term as well.

- **POS+SSR**: Probability-ordered search including search space reduction. This scheme incorporates the POS approach. On top of that, search positions are excluded that possess distortion values that are identical with the distortion values of other positions with a smaller number of bits for the corresponding motion vector.

- **NOS+SSR**: Norm-ordered search including search space reduction. Similarly operated as POS+SSR, only that the probability-ordered search is replaced by the norm-ordered search.

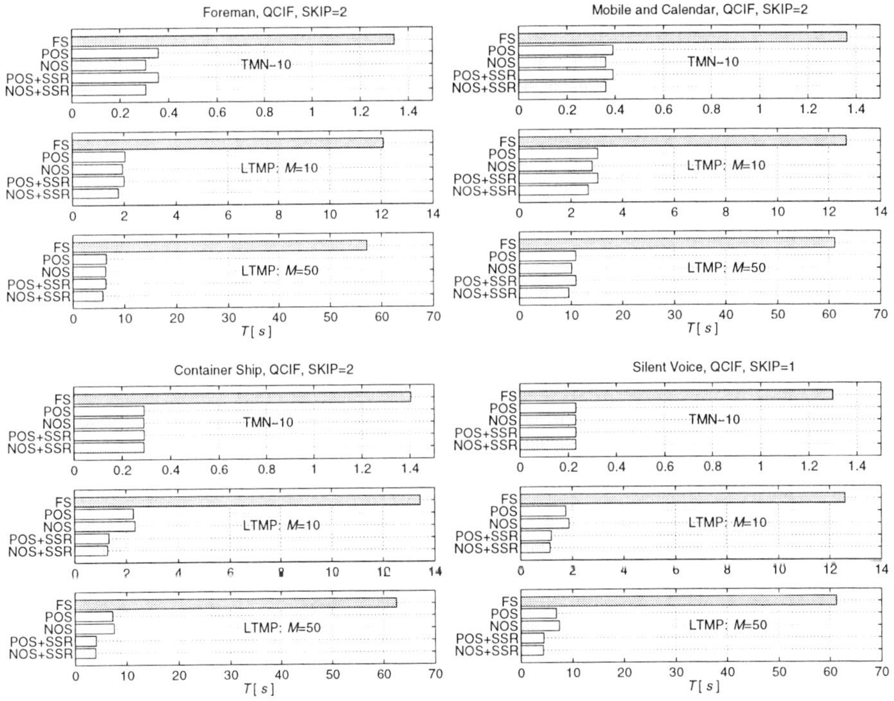

Figure 5.1. Average computation time in seconds per frame for various search strategies.

To quantify the reduction in average computation time, a factor Γ is used that describes the ratio between the average computation time of the FS method and the considered approach. More precisely, Γ is defined as

$$\Gamma = \frac{T(\text{FS})}{T(\text{P})} \tag{5.11}$$

with $T(\text{FS})$ being the average computation time for the method FS for each sequence and value of M. The computation times $T(\text{FS})$ for the FS methods will serve as the anchor for all computation time measurements in this chapter. The value $T(\text{P})$ corresponds to the average computation time that is measured for the proposed method "P".

The largest gain is obtained for the POS and NOS approaches compared to FS. This is because of the use of the triangle inequality test in (5.3). Considering the POS method for the TMN-10 coder, the largest Γ value of 5.6 is achieved for the sequence *Silent Voice* with $T(\text{FS}) = 1.30$ seconds and $T(\text{POS}) = 0.23$ seconds. The smallest Γ value of 3.5 is measured for the sequence *Mobile & Calendar* with $T(\text{FS}) = 1.36$ seconds and $T(\text{POS}) = 0.39$ seconds. The results for the NOS method are very similar to those for the POS method for the TMN-10 coder.

When employing $M = 10$ reference pictures, Γ is increased and its value is between 4.5 and 6.7. For $M = 50$ reference pictures, Γ is between 6.1 and 9.0. The search order, i.e., POS or NOS, has a relatively small impact on resulting average computation time for small values of M or for sequences with a static background like *Container Ship* and *Silent Voice* as depicted in the lower two plots of Fig. 5.1. Another important observation is that the relative gains are larger as the number of reference frames increase. For the FS case, the average computation time is roughly constant for the various sequences and proportional to the number of reference frames.

A significant reduction in average computation time can be achieved for the NOS method for large M and sequences with global motion, like *Foreman* and *Mobile & Calendar* as shown in the upper two plots of Fig. 5.1. This benefit for the NOS case over POS also extends to the combination with SSR. While the reduction of the search space provides only slight benefits for sequences with global motion, the concept seems to work well for sequences with a static background as can be seen for *Container Ship* and *Silent Voice*. The resulting values of Γ are between 4.7 and 11.1 for $M = 10$ and 6.4 and 15.9 for $M = 50$. Please note that the search space reduction as described in Section 5.1.3, does not affect the average computation time for the TMN-10 coder.

The results of this experiment lead to the conclusion that the use of the triangle inequality is beneficial when being incorporated either into POS or NOS, with a slight benefit for the NOS approach. The combination with SSR provides only small gains for sequences with global motion but significant gains for sequences with a static background. Hence, the approach that is proposed here is NOS+SSR which is employed in the next set of experiments.

Figure 5.2 presents the comparison of the average computation time when stopping at different levels in the hierarchy of triangle inequalities as described in Section 5.1.1. The time for the pre-computation step is incorporated as well. The following cases are compared

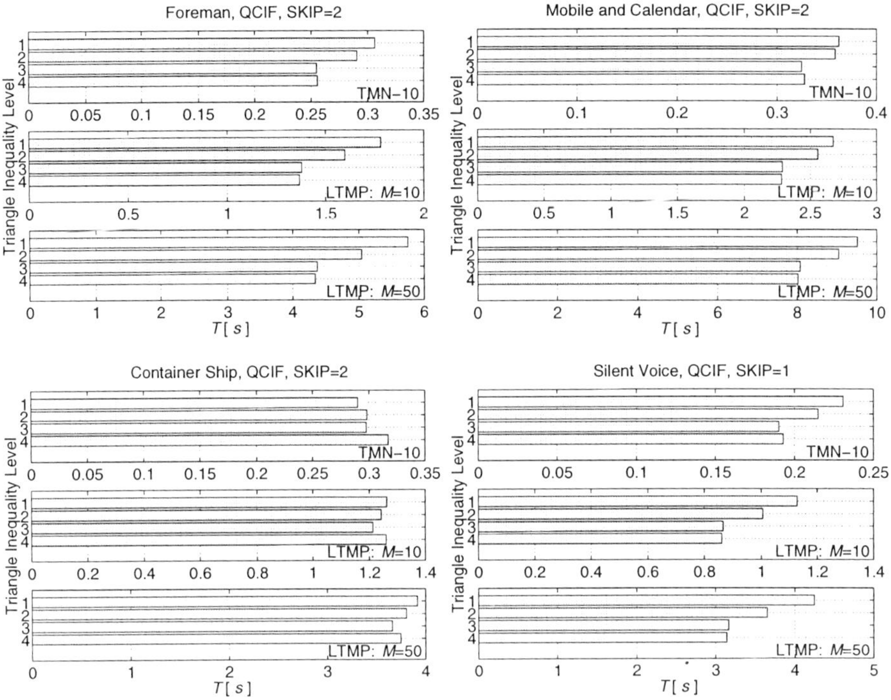

Figure 5.2. Average Computation time in seconds per frame vs. number of triangle inequalities.

- **Level 1**: only 1 triangle inequality per 16 × 16 and 8 × 8 block is tested before computing the SAD. All results for varying triangle inequality levels are based on the NOS+SSR scheme. Hence, this result is identical with the NOS+SSR result in Fig. 5.1.
- **Level 2**: in addition to the triangle inequalities that are evaluated for the level 1 case, also the 4 triangle inequalities using the norms for the 8 × 8 sub-blocks of the 16 × 16 block are evaluated. Here, no extra computation is needed on the pre-computation side, since the 8 × 8 block norms have been already computed for the level 1 case.
- **Level 3**: in addition to the triangle inequalities that are evaluated for the level 2 case, also the triangle inequalities for sub-blocks with 4 × 4 pixels are evaluated for both, 16 × 16 and 8 × 8 blocks.
- **Level 4**: in addition to the triangle inequalities that are evaluated for the level 3 case, also the triangle inequalities for sub-blocks with 2 × 2 pixels are evaluated for both, 16 × 16 and 8 × 8 blocks.

The additional benefits for using more than one triangle inequality level are comparably small. Moreover, for the sequence *Container Ship*, no reduction or even an increase in average computation time can be observed. Please note that the evaluation of triangle inequalities introduces an overhead in computation time for positions that are not excluded. Since the hierarchy of triangle inequalities is fixed for all blocks, this overhead can only be compensated if the evaluation in the hierarchy of triangle inequalities often yields an early termination of the distortion computation. Moreover, the pre-computation of the norm values requires computation time as well. This indicates why almost no reduction in computation time is measured for the sequence *Container Ship*. But, for the other sequences significant reductions in average computation time are obtained. For the TMN-10 coder, the factor of average computation time reduction against FS, Γ is between 4.1 and 6.8. For long-term memory MCP, with $M = 10$ reference frames, the Γ values are measured between 5.6 and 14.6, for $M = 50$ reference frames, the Γ values are measured between 7.6 and 19.5. Again, the Γ values increase as the number of reference frames increases. Thus, the case of 4 levels of triangle inequalities is employed in the next experiments.

5.3.2 RESULTS FOR LOSSY METHODS

The reductions in average computation time for the lossless methods are significant, but might not be sufficient in some applications. In this section, lossy methods are investigated in order to further reduce computation time while accepting small losses in rate-distortion performance. To quantify the losses in rate-distortion performance, the bit-rate that is needed to represent the video signal at an average PSNR of 34 dB is measured for the proposed method. Then, the relative bit-rate in % is computed against the bit-rate that the (lossless) TMN-10 coder transmits to represent the video sequence at 34 dB. Positive values for the relative bit-rate correspond to bit-rate savings against TMN-10 in %, while negative values indicate an increase of the bit-rate.

First, it is investigated whether the removal of blocks that are "too similar," can provide an efficient trade-off between complexity and rate-distortion performance. Figure 5.3 shows the result for the four test sequences. For the results in Fig. 5.3, the approach as described in Section 5.2.1 is employed when varying the parameter Θ. The case $\Theta = 0$ is identical with the lossless approach in Fig. 5.2 with 4 triangle inequality levels. For the other cases where Θ is varied over the values 25, 100, and 400, those search positions are removed when their corresponding measure in (5.8) falls below Θ.

For the sequences with a static background, *Container Ship* and *Silent Voice*, the variation of Θ provides very substantial reductions in average computation time for the long-term memory MCP coder which is indicated by $M = 10$ and $M = 50$ in Fig. 5.3. Considering the case of $\Theta = 25$, Γ values between 5.8 and 27 are measured for $M = 10$ reference frames, while for $M = 50$ reference frames, Γ values between 7.9 and 53 are measured. These increases of Γ values are achieved by accepting small losses in rate-distortion performance.

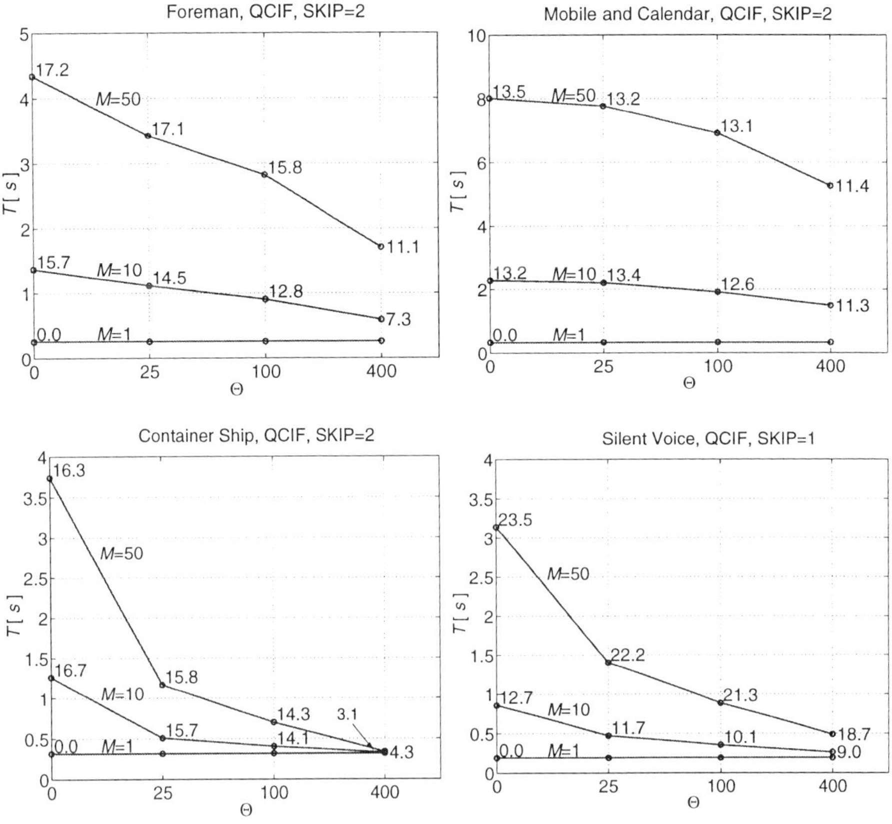

Figure 5.3. Average computation time in seconds per frame vs. threshold Θ to reduce the search range for a varying number of reference frames $M = 1, 10, 50$. The numbers in each picture indicate the bit-rate savings in % against the TMN-10 coder for the corresponding values of Θ and M.

For example, let us consider the sequence *Silent Voice* for the case of long-term memory MCP with $M = 50$ reference frames. When comparing the case $\Theta = 0$ with $\Theta = 25$, the computation time can be reduced from 3.2 seconds per frame to 1.4 seconds per frame, while the bit-rate savings against TMN-10 are reduced from 23.5 % to 22.2 %. Further reductions in average computation time for the cases $\Theta = 100$ and $\Theta = 400$ are paid by larger losses in rate-distortion performance. The reductions in average computation time for the sequences with global motion (*Foreman* and *Mobile & Calendar*) are smaller than for the sequences with a static background. But for all sequences tested here, a value of $\Theta = 25$ appears to be a good trade-off and will be employed in the next experiment.

Another option to reduce average computation time is to avoid the evaluation of blocks that are "too different," as described in Section 5.2.1. The results for

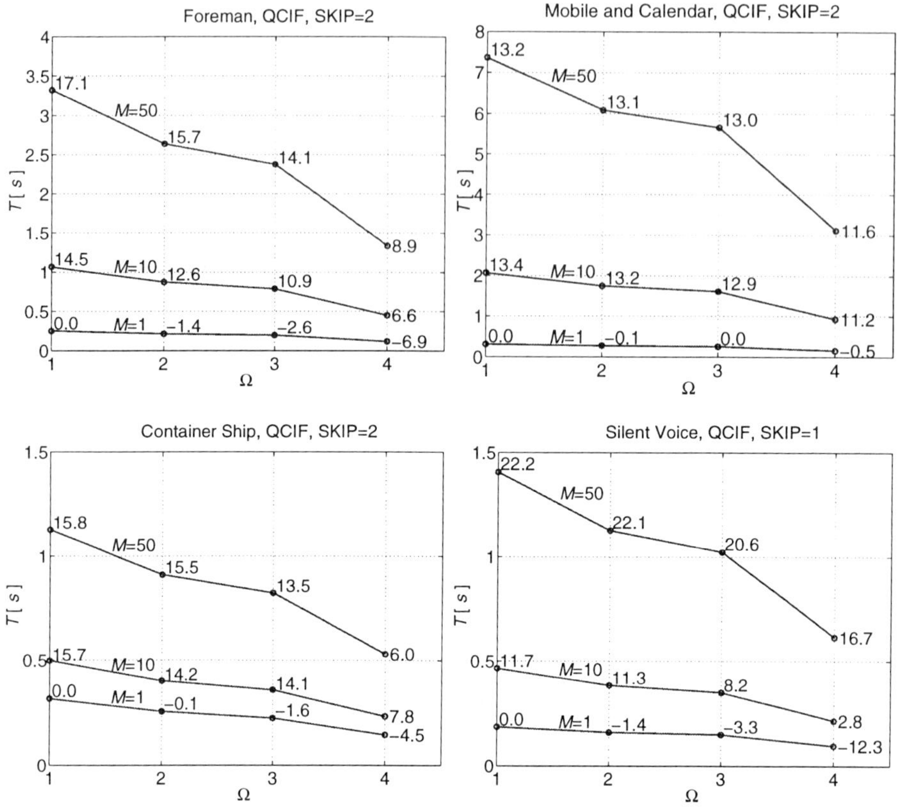

Figure 5.4. Average computation time in seconds per frame vs. Ω for early search termination for a varying number of reference frames $M = 1, 10, 50$. The numbers in each picture indicate the bit-rate savings in % against the TMN-10 coder for the corresponding values of M and Ω.

the four test sequences are shown in Fig. 5.4. For that, the parameter Ω in (5.9) is varied over the values $1, 2, 3$, and 4. The case $\Omega = 1$ is identical with the case $\Theta = 25$ in Fig. 5.3. The results for the other values of Ω in Fig. 5.4 only differ from the case $\Omega = 1$ in more positions being excluded using (5.9). The relative reductions in average computation time are similar for all sequences. Considering the case of $\Omega = 2$, the values of Γ are between 4.9 and 7.6 for $M = 1$ reference picture, between 6.8 and 33 for $M = 10$, and between 9.4 and 67 for $M = 50$ reference frames. The increase of the Γ value results in reduced rate-distortion performance. For example, when comparing $\Omega = 1$ with $\Omega = 2$ for the sequence *Silent Voice*, the average computation when using $M = 50$ reference pictures reduces from 1.4 to 1.1 seconds per frame. The bit-rate savings against TMN-10 are reduced from 23.5 to 22.1 %. Increasing Ω provides further reductions in average computation time since fewer candidates are evaluated in the search space. But also the rate-distortion performance degrades. Thus, the choice of $\Omega = 2$ will be employed in the last experiment of this chapter.

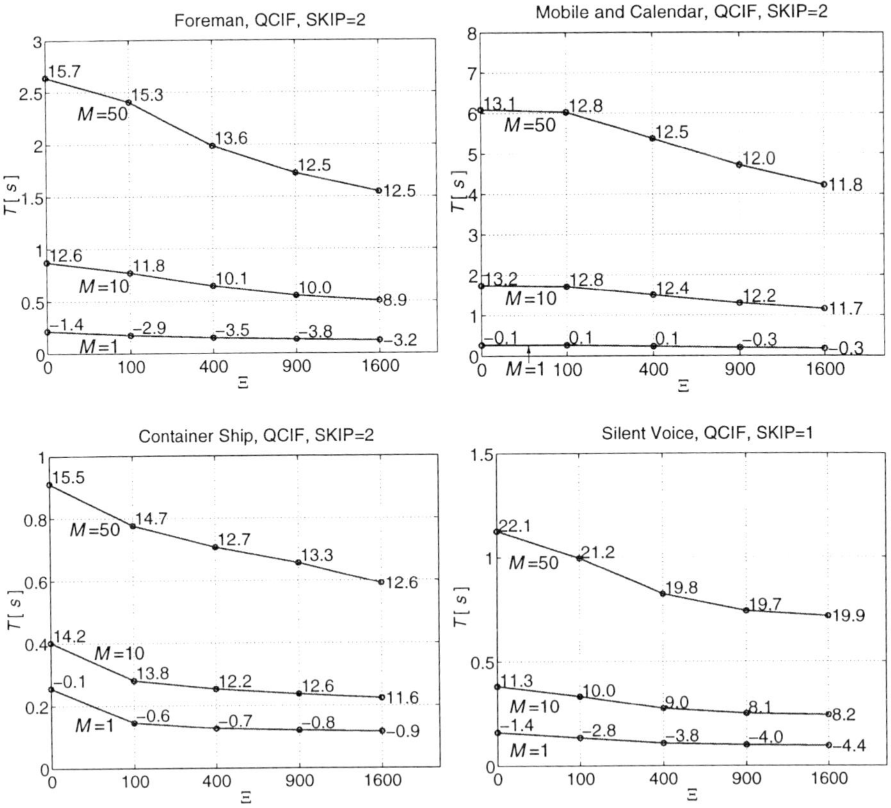

Figure 5.5. Average computation time in seconds per frame vs. Ξ for reduced resolution distortion computation for a varying number of reference frames $M = 1, 10, 50$. The numbers in each picture indicate the bit-rate savings against the TMN-10 coder for the corresponding values of M and Ξ.

Finally, the achieved reduction in average computation time is measured, when sub-sampling the block for which distortion is computed. As described in Section 5.2.2, this approach is incorporated by stopping the distortion evaluation at a certain level in the hierarchy of triangle inequalities. The results in Fig. 5.5 are obtained by varying the threshold Ξ over the values $0, 100, 400, 900,$ and 1600 when applying the inequality (5.10) to decide whether to compute distortion on the pixel or 2×2 triangle inequality level. The case $\Xi = 0$ is identical to the case $\Omega = 2$ in Fig. 5.4. Considering the case of $\Xi = 100$, a further increase of Γ can be achieved. The resulting values of Γ are between 4.6 and 9.3 for $M = 1$ reference frame, between 6.7 and 46 for $M = 10$, and between 10 and 77 for $M = 50$ reference frames. The increase in Γ again results in reduced rate-distortion performance. For example, when comparing the results for the sequence *Silent Voice* for $\Xi = 1$ and $\Xi = 100$, the average computation when using $M = 50$ reference pictures reduces from 1.1 to 1.0 seconds per frame. The bit-rate savings against TMN-10 are reduced from 23.5 to 21.2 %.

5.4 DISCUSSION AND OUTLOOK

The experiments for the proposed approaches to computation time reduction show that the proposed methods provide significant speed-ups. But those reductions are dependent on the sequence. For the sequences *Container Ship* or *Silent Voice*, the reductions in average computation time are large. On the other hand, for the sequence *Mobile & Calendar*, the presented concepts do not provide such drastic reductions in average computation time. This varying behavior may be caused by the choice of the parameters for the various approaches, especially for the results of the lossy methods. Thus, further improvements could be achieved, if the parameter choices are adapted to the various sequences.

The techniques presented, provide especially large reductions in average computation time for increasing memory sizes. Among the presented approaches, the reduction of the search space via the exclusion of identical or similar blocks based on frame differences is especially successful for sequences with static background. Note that also other methods could be used to identify identical or similar blocks in the search space, such as following the trajectory of integer-pixel motion vectors similar to the Error Tracking approach in [FSG96, SFG97, GF99, FGV98]. However, these methods might become quite complex and the number of removed positions might become too small compared to the additional overhead. Moreover, the rate to indicate such a shifted position might be smaller in the older frame, when the corresponding match is found for an adjacent block. Thus, since a Lagrangian cost function is minimized it is not always the case that the position in the older frame corresponds to higher costs.

Another issue is the potential reduction of bit-rate due to the exclusion of search positions. This is possible because shorter code words can be used to indicate the remaining positions. This idea might become important for steady background, which usually shows only small gains in prediction performance and with that justifies only a small bit-rate overhead. But this idea incurs an error resilience problem, since the decoder has to decide which pixels to exclude simultaneously. Thus, if a transmission error occurs, this decision at the decoder might be different from the encoder and therefore the picture reference parameters get out of synchronization.

5.5 CHAPTER SUMMARY

In this chapter, it has been demonstrated that the computation requirements for multi-frame motion estimation can be reduced by more than an order of magnitude, while maintaining all or most of the improvements in coding efficiency. The multi-frame scenario considered is long-term memory MCP employing a sufficiently large number of reference frames ($M \geq 10$). When a picture is decoded and included into the set of M reference pictures, the fact is exploited that $M - 1$ of the pictures and the associated pre-computed data remain unchanged in the multi-frame buffer. Another important fact that is exploited is that many blocks in the multi-frame buffer are quite similar and can thus be excluded from the search space. These aspects distinguish the fast search methods for multi-frame MCP from most single-frame approaches and provide the large reductions in average computation time.

To illustrate the impact of the various techniques, experiments are conducted, which are evaluated using a speed-up factor Γ that specifies the ratio between the average computation time of the anchor search technique and a proposed method. Without incurring any loss in rate-distortion performance, a value of Γ between 4.1 and 6.8 for the TMN-10 coder can be achieved. For long-term memory MCP, Γ is measured between 5.6 and 14.6 for $M = 10$ and for $M = 50$ reference frames and Γ value between 7.6 and 19.5 can be achieved. Larger gains can be achieved when accepting small losses in rate-distortion performance. These losses against the lossless anchor method are quantified by the corresponding bit-rate reduction at a reconstruction level of 34 dB in PSNR. When accepting small losses in rate-distortion performance for the long-term memory MCP coder, Γ values up to 46 can be achieved for $M = 10$ reference frames, and for $M = 50$ frames a speed-up by a factor of 77 is reported. Hence, the methods that are presented in this chapter show, that the problem of increased computational complexity for multi-frame motion estimation is practically tractable.

Chapter 6

ERROR RESILIENT VIDEO TRANSMISSION

Multi-frame MCP improves the rate-distortion efficiency of video codecs in the case of error-free transmission of the bit-stream. In this chapter, the efficiency of long-term memory MCP for a transmission scenario where the bit-stream may be received in error is investigated. The most important applications today including wireless packet networks show burst errors. The complete removal of such burst errors using channel coding techniques is very costly in terms of overhead bit-rate when assuming a limited end-to-end delay. Hence, source coding techniques are employed to improve error resilience while allowing some transmission errors to occur.

A video signal that is compressed using a hybrid video coder is extremely vulnerable to transmission errors. When the bit-stream is received in error, the decoder cannot or should not reconstruct the affected parts of the current frame. Rather a concealment is invoked. But motion compensation in combination with concealed image content leads to inter-frame error propagation which causes that the reference frames at encoder and decoder differ. In this scenario random reconstruction results are produced at the decoder side which depend on the statistics of the transmission errors that cause a concealment and the motion compensation that determines the inter-frame error propagation. The reconstruction quality at the decoder is determined by the source coding distortion, which quantifies the error between the original signal and the reconstructed signal at the encoder, and the expected transmission error distortion, which quantifies the error between the reconstructed signals at encoder and decoder.

The task of the coder control is to determine the coding parameters so as to optimize the rate-distortion performance at the decoder. For long-term memory MCP, inter-frame error propagation has to be considered in the multi-frame buffer which is affected by the choice of the picture reference parameter. Therefore, the novel coder control in this chapter incorporates the estimate of the expected transmission error distortion into the Lagrangian cost functions for the selection of the motion vectors including the picture reference parameter and the macroblock modes. Experimental results with a Rayleigh fading channel show that long-term memory MCP significantly outperforms single-frame MCP in the presence of channel errors.

This chapter is organized as follows. In Section 6.1, the extensions of the video decoder for error resilient transmission are described. The proposed coder control is explained in Section 6.2. First, the effect of inter-frame error propagation is illustrated. Second, the estimation of the expected transmission error distortion is presented. Finally, the incorporation of the expected transmission error distortion into Lagrangian coder control is described. Section 6.3 presents experimental results that evaluate the new approach for transmission scenarios without and with feedback.

6.1 ERROR RESILIENT EXTENSIONS OF THE DECODER

The video decoder and syntax employed in this chapter are similar to the long-term memory codec in Chapter 3. On top of that, the decoder is extended to cope with transmission errors because the channel coder is not expected to correct all errors.

The multiplexed video bit-stream consists of variable length code words. Hence, a single bit error may cause a loss of synchronization and a series of erroneous code words at the decoder. The common solution to this problem is to insert unique synchronization code words into the bit-stream in regular intervals, usually followed by a block of "header" bits. The H.263 standard supports optional GOB-headers as re-synchronization points which are also used throughout this chapter. A GOB in QCIF format usually consists of 11 macroblocks that are arranged in one row. Because all information within a correctly decoded GOB can be used independently from previous information in the same frame, the GOB is often used as the basic unit for decoding. Hence, if a transmission error is detected, the GOB is discarded entirely.

The severeness of the error caused by discarded GOBs can be reduced if error concealment techniques are employed to hide visible distortion as much as possible. In the simulation environment of this work, the simple and most common approach called *previous frame concealment* is employed, i.e., the corrupted image content is replaced by corresponding pixels from the previous frame. This is conducted by setting the macroblocks in the discarded GOB to the SKIP mode. The concealment scheme can be applied simultaneously at decoder and encoder yielding the same result at both ends. This simple approach yields good concealment results for sequences with little motion [Che95]. However, severe distortions may be introduced for image regions containing a large amount of motion.

6.2 ERROR-RESILIENT CODER CONTROL

Given the decoder and the error resilient extensions as described in the previous section, the task of the coder control is to determine the coding parameters that generate a bit-stream which optimizes reconstruction quality at the decoder. In this work, the quality at the decoder for single-frame and long-term memory MCP is compared for a fixed overall transmission rate. Ideally, all transmission parameters including source coding, channel coding, and packetization should be controlled to provide a meaningful comparison for a given channel. The latter, packetization, is considered to have a similar effect on both methods, single-frame and long-term memory MCP, and is therefore fixed for the comparisons. However, channel coding and source coding must be considered jointly since the parameter choices for both mutually affect each other and fixing one of them might bias the comparison between single-frame and long-term memory MCP.

Channel coding, which is realized in this work by forward error correction (FEC) techniques, directly affects the number of transmission errors and with that the concealment energy. An increasing number of bits to correct channel errors reduces the likelihood of concealed image content but decreases the bit-rate for the video source coder. The parameters of the video coder affect the source coding distortion and the amount of inter-frame error propagation. Hence, the problem of bit-allocation for the transmission system should ideally be controlled by this trade-off, which is addressed in this work by a constrained optimization approach. For that, the number of bits for FEC and with that the bit-rate for the source coder is constrained over the entire sequence. The number of bits for FEC determines the likelihood of transmission errors and together with the motion in the scene affects the average concealment energy. The average concealment energy is therefore assumed to be fixed for the entire sequence as well. What remains is to control the trade-off between source coding distortion and inter-frame error propagation. These two quantities are affected by the choice of the motion vectors including the picture reference parameter and the macroblock modes. The optimal transmission result is finally obtained by picking the number of bits for FEC which correspond to the highest average decoder PSNR.

The remainder of this section discusses the control of the trade-off between source coding distortion and inter-frame error propagation. In Section 6.2.1, the effect of inter-frame error propagation is illustrated. Then, in Section 6.2.2, the approach to the estimation of the expected transmission error distortion is presented. The trade-off between source coding distortion and inter-frame error propagation is achieved by incorporating the estimate of expected transmission error distortion into the Lagrangian coder control as presented in Section 6.2.3.

6.2.1 INTER-FRAME ERROR PROPAGATION

When a transmission error occurs, the corresponding GOBs are lost and concealed using previous frame concealment as described above. Hence, the reconstructed frames at encoder and decoder differ. Referencing this image content for MCP leads to inter-frame error propagation. In the following, it is assumed that the nine GOBs of each QCIF picture are transmitted in one packet and each packet is lost with probability p or correctly received with probability $q = 1 - p$. Although the transmission scenario in the experiments that are described later does not employ such a packetization scheme, the assumption that one picture is transmitted in one packet greatly simplifies the analysis here. Moreover, the approximation of the expected transmission error distortion as presented in the next section relies on this assumption.

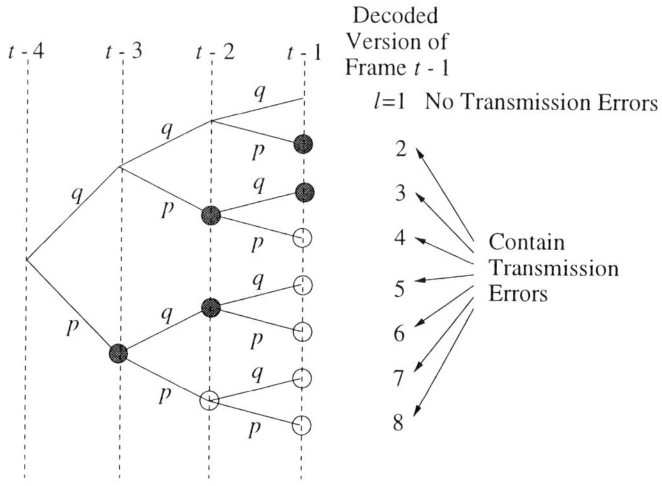

Figure 6.1. Binary tree of possible error events. Each node of the tree corresponds to a decoded version of a video frame. The nodes labeled with a circle are those that contain transmission errors. The shaded circles correspond to the error cases considered.

Figure 6.1 illustrates the combination of possible error events in case of single-frame MCP, i.e., only the prior decoded frame can be referenced for motion compensation. Let us assume, that the frame at time instant t that references frame $t-1$ is currently being coded. The goal is to estimate the average errors that have accumulated in frame $t-1$ to incorporate these measures into the coder control. For that, frames older than frame $t-1$ have to be considered due to inter-frame error propagation. For the sake of simplicity, it is assumed that the frame at time instant $t - 4$ is correctly decoded. In the next frame at time instant $t - 3$, reference is made to frame $t - 4$ using motion compensation. The image content at time instant $t - 3$ is either lost and concealed with probability p or correctly decoded with probability $q = 1 - p$. Hence, the two nodes at time instant $t - 3$ correspond to two decoded versions of that video frame. The decoding of image content in the frame at time instant $t - 2$ which references frame $t - 3$ results in 4 combinations of possible outcomes while image content in the frame at time instant $t - 1$ can be decoded in 8 different ways. It is easy to conclude that any succeeding frame doubles the number of possibilities of the decoding result. Hence, modeling all these branches of the event tree would very quickly be intractable since $L = 2^t$ combinations would have to be computed for a frame that is t time instants decoded after the first frame.

If long-term memory MCP is utilized, the number of branches leaving a node in the tree of possible error events varies since frames other than just the prior decoded frame can also be referenced. Moreover, since each macroblock or block can reference a different picture in the multi-frame buffer, the tree of possible error events has to be used for each pixel. Pixel resolution is required since the spatial displacement may cause that the MCP signal covers block boundaries in the reference picture and those blocks might reference different pictures. This results in $L = 2^t$ combinations per pixel after t time instants.

6.2.2 ESTIMATION OF THE EXPECTED TRANSMISSION ERROR DISTORTION

The illustrations in the previous section provides the insight that losses of pictures cause a concealment of image content which leads to inter-frame error propagation. Inter-frame error propagation leads to different decoded pictures that correspond to the branches of the binary tree in Fig. 6.1. Given the random decoding results, the task of the coder control is to choose the encoding parameters so as to optimize the average rate-distortion performance at the decoder. Hence, the distortion at the decoder becomes a random variable and we have to determine a relationship between the expected values of the distortion random variable and the coder control parameters. Encoding proceeds in two steps, which are motion estimation and macroblock mode decision. Hence, we first compute the expected distortion $E\left\{\mathcal{D}_{\text{DFD}}\right\}$ of the random variable $\mathcal{D}_{\text{DFD}}$ that is associated with the MCP error at the decoder to conduct motion estimation. Second, the expected distortion $E\left\{\mathcal{D}_{\text{REC}}\right\}$ that corresponds to the reconstruction error is computed for macroblock mode decision.

Let us assume that at time instant $t - m_t$ the decoded picture is $\acute{s}_l\left[x, y, t - m_t\right]$ with probability p_l. The expected distortion $E\left\{\mathcal{D}_{\text{DFD}}(\boldsymbol{S}_i, \boldsymbol{m})\right\}$ that is associated to the MCP error for the block $\boldsymbol{S}_i$ when utilizing the motion vector $\boldsymbol{m}$ including spatial displacements (m_x, m_y) and picture reference parameter m_t is computed as

$$E\{\mathcal{D}_{\text{DFD}}(\boldsymbol{S}_i,\boldsymbol{m})\} = \sum_{l=1}^{L} p_l D_{\text{DFD}}^l(\boldsymbol{S}_i,\boldsymbol{m}) \tag{6.1}$$

$$= \sum_{l=1}^{L} p_l \sum_{(x,y)\in\boldsymbol{\mathcal{A}}_i} (s[x,y,t] - \acute{s}_l[x-m_x, y-m_y, t-m_t])^2,$$

with $\boldsymbol{\mathcal{A}}_i$ being the set of pixels in the block $\boldsymbol{S_i}$. Note that the L distortion terms D_{DFD}^l correspond to each different outcome of the decoding at time instant $t - m_t$ and have to be evaluated for each position in the search space. This is computationally very demanding and the following approximation is proposed.

Let the reference frame in the lth decoding branch be expressed by the correctly decoded reference frame $\acute{s} = \acute{s}_1$ plus a remaining error v_l

$$\acute{s}_l[x,y,t-m_t] = \acute{s}[x,y,t-m_t] + v_l[x,y,t-m_t]. \tag{6.2}$$

The distortion term D_{DFD}^l is approximated by

$$D_{\text{DFD}}^l(\boldsymbol{S}_i,\boldsymbol{m}) = \sum_{(x,y)\in\boldsymbol{\mathcal{A}}_i} (s[x,y,t] - \acute{s}_l[x-m_x,y-m_y,t-m_t])^2 \tag{6.3}$$

$$= \sum_{(x,y)\in\boldsymbol{\mathcal{A}}_i} (s[x,y,t] - \acute{s}[x-m_x,y-m_y,t-m_t] - v_l[x-m_x,y-m_y,t-m_t])^2$$

$$\approx \sum_{(x,y)\in\boldsymbol{\mathcal{A}}_i} (s[x,y,t] - \acute{s}[x-m_x,y-m_y,t-m_t])^2 + v_l^2[x-m_x,y-m_y,t-m_t],$$

where the cross terms

$$\sum_{(x,y)\in\boldsymbol{\mathcal{A}}_i} s[x,y,t]\cdot v_l[x-m_x,y-m_y,t-m_t] \approx 0 \qquad \text{and} \tag{6.4}$$

$$\sum_{(x,y)\in\boldsymbol{\mathcal{A}}_i} \acute{s}[x-m_x,y-m_y,t-m_t]\cdot v_l[x-m_x,y-m_y,t-m_t] \approx 0$$

are neglected since s and $\acute{s}$ are assumed to be uncorrelated from v_l and v_l is assumed to have a zero mean value. The expected distortion that is associated to the MCP error is approximated by

$$E\{\mathcal{D}_{\text{DFD}}(\boldsymbol{S}_i,\boldsymbol{m})\} \approx \sum_{(x,y)\in\boldsymbol{\mathcal{A}}_i} (s[x,y,t] - \acute{s}[x-m_x,y-m_y,t-m_t])^2$$

$$+ \sum_{l=2}^{L} p_l \sum_{(x,y)\in\boldsymbol{\mathcal{A}}_i} v_l^2[x-m_x,y-m_y,t-m_t] \tag{6.5}$$

with

$$\sum_{l=1}^{L} p_l = 1. \tag{6.6}$$

Note that the first term on the right-hand side in (6.5) corresponds to the distortion term computed for motion estimation in error-free transmission (D_{DFD}). The second term on the right-hand side in (6.5) represents the estimate of the expected transmission error distortion.

In the following, expressions for the approximation of the expected distortion at the decoder for the macroblock modes INTRA, INTER, INTER+4V, and SKIP are derived. For the INTRA macroblock mode, the associated distortion values at coder and decoder are identical since no MCP is conducted. For the INTER macroblock mode, the expected distortion $E\{\mathcal{D}_{\text{REC}}\}$ that corresponds to the reconstruction error can be derived similar to distortion $E\{\mathcal{D}_{\text{DFD}}\}$. Given the motion vector $\boldsymbol{m}_i$ that has been chosen in the preceding motion estimation step for the block $\boldsymbol{S}_i$, the average distortion $E\{\mathcal{D}_{\text{REC}}(\boldsymbol{S}_i, \text{INTER}|Q, \boldsymbol{m}_i)\}$ is given as

$$E\{\mathcal{D}_{\text{REC}}(\boldsymbol{S}_i, \text{INTER}|Q, \boldsymbol{m}_i)\} = \sum_{l=1}^{L} p_l D^l_{\text{REC}}(\boldsymbol{S}_i, \text{INTER}|Q, \boldsymbol{m}_i), \tag{6.7}$$

with

$$D^l_{\text{REC}}(\boldsymbol{S}_i, \text{INTER}|Q, \boldsymbol{m}_i) = \sum_{(x,y)\in\boldsymbol{\mathcal{A}}_i} (s[x,y,t] - \acute{s}_l[x - m_{ix}, y - m_{iy}, t - m_{it}] - \acute{u}[x,y,t])^2. \tag{6.8}$$

The signal $\acute{u}$ corresponds to the coded MCP error signal when utilizing the DCT quantizer value Q. With (6.2) and (6.6), the expected distortion that is associated to the reconstruction error is approximated by

$$E\{\mathcal{D}_{\text{REC}}(\boldsymbol{S}_i, \text{INTER}|Q, \boldsymbol{m}_i)\} \approx \tag{6.9}$$

$$\sum_{(x,y)\in\boldsymbol{\mathcal{A}}_i} (s[x,y,t] - \acute{s}[x - m_{ix}, y - m_{iy}, t - m_{it}] - \acute{u}[x,y,t])^2$$

$$+ \sum_{l=2}^{L} p_l \sum_{(x,y)\in\boldsymbol{\mathcal{A}}_i} v_l^2[x - m_{ix}, y - m_{iy}, t - m_{it}], \tag{6.10}$$

where in addition to the cross terms in (6.4) the cross term

$$\sum_{(x,y)\in\boldsymbol{\mathcal{A}}_i} \acute{u}[x,y,t] \cdot v_l[x - m_{ix}, y - m_{iy}, t - m_{it}] \approx 0 \tag{6.11}$$

is neglected as well with the assumption that $\acute{u}$ and v_l are uncorrelated and both having a zero mean value. The first term on the right-hand side in (6.10) corresponds to the distortion term that is computed for the INTER macroblock mode in error-free transmission (D_{REC}). The second term on the right-hand side in (6.10) is the expected transmission error distortion and identical to the second term on the right-hand side in (6.5). Employing similar arguments, the expected distortions at the decoder for the INTER+4V and SKIP macroblock modes can be derived.

With the assumptions in (6.4) and (6.11), the expressions for the expected distortion for motion estimation and macroblock mode decision are greatly simplified, since the current original video signal s is decoupled from the computation of the estimate for the expected transmission error distortion. Nevertheless, the computational burden is still very high because of the large number of combinations involved to obtain the value for the estimate of the expected transmission error distortion. Hence, the number of possibilities of different decoded reference pictures is restricted to two cases:

1. The referenced image content is in error and concealed, (branch $l = 2$ in Fig. 6.1).

2. The referenced image content has been correctly decoded but references concealed image content (branch $l = 3$ in Fig. 6.1).

In Fig. 6.1, each node of the tree corresponds to a decoded version of a video frame. The nodes labeled with a circle are those that contain transmission errors. The approximation incorporates only those cases with shaded circles. This approximation is justified by assuming p to be very small and two error events in a row to be very unlikely. Other decoded versions are neglected assuming that if an error has occurred several time instants in the past, it has then several times been motion-compensated and therefore filtered and somewhat reduced. Nevertheless, these assumptions may not hold for some cases. Hence, possible shortcomings of this approximation are analyzed in the next section using experimental results when comparing to a more accurate estimate for the expected transmission error distortion.

In practice, the computation of the expected transmission error distortion is conducted via re-decoding each transmitted picture by employing the corresponding concealed reference frames. The decoding has to be done for each of the considered nodes, i.e., twice per frame. The big advantage of this approach is that it includes effects like overlapped block motion compensation and de-blocking filters. Moreover, it can be used similarly for single-frame and long-term memory MCP. Finally, each of the two results of the decoding ($\acute{s}_2$ and $\acute{s}_3$) is subtracted from the correctly decoded signal $\acute{s}$. The resulting difference signals (v_2 and v_2) are combined as follows

$$D_{\text{ERR}}(\boldsymbol{S}_i, \boldsymbol{m}) = \sum_{l=2}^{3} \sum_{(x,y) \in \boldsymbol{\mathcal{A}}_i} v_l^2[x - m_x, y - m_y, t - m_t], \tag{6.12}$$

to arrive at the approximate *error modeling* term for consideration of the transmission errors. The values of D_{ERR} can be efficiently pre-computed utilizing an algorithm similar to the one proposed in [LS95]. In (6.12) the probability weight is left out because the two considered events are equally likely in the error event tree. The weighting is addressed in the next section when the transmission error modeling term is incorporated into the Lagrangian coder control.

In this work, cases without and with feedback are considered. If no feedback is available, the error modeling term in (6.12) is computed only once and then attached to each reference frame for incorporation into the Lagrangian cost functions for the encoding of later frames. In case feedback messages are sent from the decoder to the encoder, the error modeling term in (6.12) is updated accordingly. Note that in this work feedback is provided about correct as well as concealed GOBs (ACK+NACK). Moreover, it is assumed that feedback messages are transmitted without error and that the exact concealment method is known to encoder and decoder. This assumption is justified by considering the relatively small bit-rate of the feedback messages which permits a relatively large amount of channel coding.

If at time instant t a feedback message is received at the encoder about a frame that was transmitted at time instant $t - \Delta t$, the encoder exactly duplicates the frame $t - \Delta t$ from the decoder using the specified concealment if needed. Then, the depending frames, i.e., frames that were transmitted later than frame $t - \Delta t$ are decoded again. Note that this decoding is only necessary for macroblocks either referencing concealed image content in frame $t - \Delta t$ or image content that is in error because of inter-frame error propagation due to concealed image content in frame $t - \Delta t$. A similar idea has been exploited in [GL94]. The error modeling is updated in that, D_{ERR} is set to 0 in the frame at time instant $t - \Delta t$ and an update of D_{ERR} is made for all depending frames.

6.2.3 INCORPORATION INTO LAGRANGIAN CODER CONTROL

The error modeling term is incorporated into motion estimation and macroblock mode decision as follows. The Lagrangian cost term of the minimization routine to determine the motion vector $\boldsymbol{m}_i$ for the block $\boldsymbol{S}_i$ in (2.8) is modified in that the weighted transmission error modeling term is incorporated

$$\boldsymbol{m}_i = \underset{\boldsymbol{m} \in \mathcal{M}}{\operatorname{argmin}} \left\{ D_{\text{DFD}}(\boldsymbol{S}_i, \boldsymbol{m}) + \kappa D_{\text{ERR}}(\boldsymbol{S}_i, \boldsymbol{m}) + \lambda_{\text{MOTION}} R_{\text{MOTION}}(\boldsymbol{S}_i, \boldsymbol{m}) \right\}, \tag{6.13}$$

where the SSD distortion measure is used for the computation of D_{DFD} and $\mathcal{M}$ contains the set of search positions. The weighting factor κ is used as a free parameter in the simulation discussed below and is necessary because it provides a means to adapt the weight of D_{ERR} to the actual concealment energy which is given by the channel conditions and the number of bits for FEC. Note that in contrast to (6.5), no error probability is included in (6.12). Hence, κ is used to scale D_{ERR} according to the effective loss probability. For a practical system it would be necessary to set κ correctly during encoding. Because in this work mainly performance bounds are investigated, the term κ is used as a free parameter and its value is chosen so as to obtain optimal overall performance (as given in maximum decoder PSNR when fixing transmission bit-rate). Also note that κ actually could be adapted on a frame or macroblock basis for improved performance. For simplicity, however, a fixed value of κ is used for a given sequence and channel.

To conduct the macroblock mode decision, the Lagrangian costs for the modes INTER, INTER+4V and SKIP in (2.7) are modified to

$$J_{\text{MODE}}(\boldsymbol{S}_k, I_k|Q, \lambda_{\text{MODE}}) = \\ D_{\text{REC}}(\boldsymbol{S}_k, I_k|Q) + \kappa D_{\text{ERR}}(\boldsymbol{S}_k, \boldsymbol{m}) + \lambda_{\text{MODE}} \cdot R_{\text{REC}}(\boldsymbol{S}_k, I_k|Q), \quad (6.14)$$

while the Lagrangian costs for INTRA mode remain unchanged as in (2.7). Note that INTRA coding terminates branches of the binary tree in Fig. 6.1, since in the case of correctly decoded image content, inter-frame error propagation is stopped. Therefore, one impact of the error modeling term when incorporated into the macroblock mode decision of the H.263 and long-term memory MCP codecs is that the number of macroblocks coded in INTRA should be increased.

The frame selection is also affected for the long-term memory MCP codec, since the error modeling term is incorporated into the Lagrangian cost function for motion estimation, which has a strong impact on the statistics of the frame selection in the case of feedback for a frame that is older than the prior coded frame. Let us assume that a feedback message is received at time instant t for a frame that was transmitted at time instant $t - \Delta t$ which is older than the prior decoded picture. As described above, the encoder can duplicate the exact reconstruction of the frame $t - \Delta t$ at the decoder and the estimate for the transmission error in (6.12) is set to $D_{\text{ERR}} = 0$. Thereby, the reference frame corresponding to time instant $t - \Delta t$ is typically referenced more frequently than the prior decoded frame (for which no feedback information is available and the error modeling term is typically $D_{\text{ERR}} \geq 0$). But, the variable length code that is used to transmit the picture reference parameter is designed for transmission scenarios without channel errors, where the frequency of selecting a reference picture is inversely proportional to the time interval between the current and the reference picture. In order to transmit the picture reference parameters for this case with the smallest possible average bit-rate, the picture reference parameters are sorted in descending order of their frequency. The result of the sorting is then transmitted to the decoder and multi-frame motion compensation is conducted by utilizing the *Index Mapping* memory control.

6.3 EXPERIMENTS

Before presenting results for the proposed framework, the transmission system including the error control channel used for the experiments is described. The description follows the basic block diagram of a video transmission system as illustrated in Fig. 6.2.

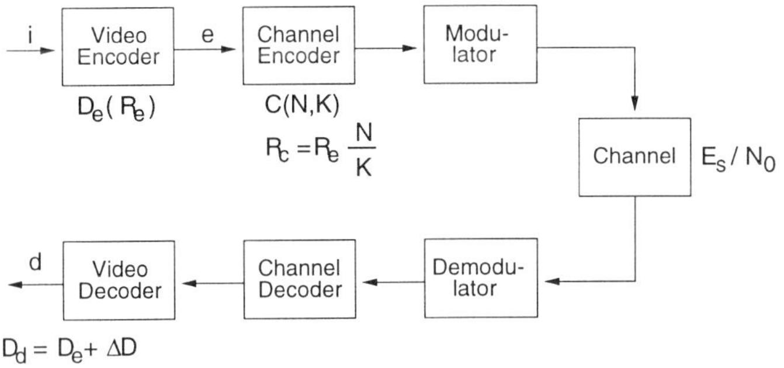

Figure 6.2. Basic components of a video transmission system.

For the channel model, modulation scheme, and channel codec, standard components are used rather than advanced techniques that reflect the current state of research. This is justified by our focus on video coding and by the fact that the selected standard components are well suited to illustrate the basic problems and trade-offs. Therefore, the described scenario should be considered as an example that is used for illustration, rather than a proposal for an optimized transmission scheme.

6.3.1 CHANNEL MODEL AND MODULATION

The simulations in this chapter are based on bit error sequences that are used within ITU-T Video Coding Experts Group for the evaluation of current and future error resilience techniques in H.263. The sequences are generated assuming Rayleigh fading with different amounts of channel interference, characterized by ratios of bit-energy to noise-spectral-energy (E_s/N_0) in the range of 14 to 30 dB.

The correlation of fading amplitudes is modeled according to the widely accepted model by Jakes [Jak74]. In this model, the correlation depends significantly on the Doppler frequency f_D, which is equal to the mobile velocity divided by the carrier wavelength. For a given carrier frequency, the correlation increases with decreasing mobile velocity, such that slowly moving terminals encounter longer burst errors. For more information on this very common channel model, see [Jak74, Skl97].

The modulation scheme and relevant parameters such as carrier frequency and modulation interval are roughly related to the ETSI standard DECT. Though

DECT is originally intended for cordless telephony, it provides a wide range of services for cordless personal communications which makes it very attractive for mobile multimedia applications [PGH95, WB95]. Similar to DECT, we use binary phase shift keying for modulation and a carrier frequency of f_c = 1900 MHz. For moderate speeds a typical Doppler frequency is f_D = 62 Hz, which will be used throughout the simulations in the remainder of this chapter. According to the double slot format of DECT, a total bit-rate of $R_t = 80$ kbit/s is assumed to be available for both source and channel coding. For simplicity, no time division multiple access structure is considered and a modulation symbol interval of $T_s = 1/80\ ms$ is used. The resulting bit error sequences exhibit severe burst errors that limit the effective use of FEC. Therefore, even at low channel code rates, residual errors cannot be avoided completely by the channel codec and have to be processed by the video decoder.

6.3.2 CHANNEL CODING AND ERROR CONTROL

For channel coding a FEC scheme that is based on RS (Reed-Solomon) codes is employed [Wic95]. For symbols composed of m bits, the encoder for an RS(N, K) code groups the incoming data stream into blocks of K information symbols (Km bits) and appends $N - K$ parity symbols to each block. Hence, the transmitted output block contains N symbols and each block is treated independently by the channel codec.

The bit-allocation between source and channel coding can be described by the *code rate* r, which is defined as $r = K/N$. For RS codes operating on m-bit symbols, the maximum block length is $N_{\max} = 2^m - 1$. By using *shortened* RS codes, any smaller value for N can be selected, which provides a great flexibility in system design. As RS codes operating on 8-bit symbols are very popular and powerful, a packet size of $N = 88$ bytes is used with $m = 8$.

An RS(N, K) decoder can correct any pattern of bit errors with less than $E < (N - K)/2$ symbols in error. In other words, for every two additional parity symbols, an additional symbol error can be corrected. If more than E symbol errors are contained in a block, the RS decoder fails and indicates an erroneous block to the video decoder. The probability that a block cannot be corrected is usually described by the RWER (residual word error rate). In general, the RWER decreases with decreasing K and/or with increasing E_s/N_0. For simplicity, the occurrence of undetected errors is ignored, whose probabilities are usually very small compared to the RWER. This is also justified by the fact, that the video decoder itself usually has some error detection capability due to syntax violations that can be exploited.

For the described channel code, modulation scheme, and channel model, this relationship is summarized in Fig. 6.3 which illustrates the RWER for the values of E_s/N_0 and K that are used in the simulations.

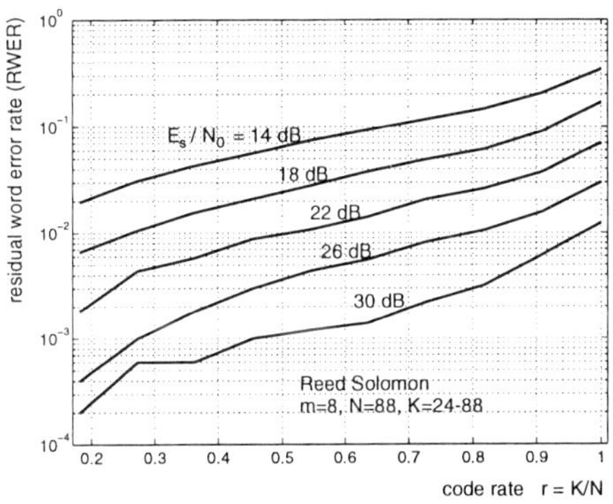

Figure 6.3. Residual word error rate (RWER) after channel coding for a Rayleigh fading channel (Doppler frequency $f_D = 62$ Hz) with Binary Phase Shift Keying.

The curves in Fig. 6.3 show that the RWER for a given value of E_s/N_0 can be reduced by approximately one order of magnitude by varying the code rate in the illustrated range. Though this reduction is already very helpful for video transmission, the observed gain in RWER is actually very moderate due to the bursty nature of the wireless channel and the limited end-to-end delay. For channels without memory, such as the AWGN channel, the same reduction in r would provide a significantly higher reduction in RWER. For the AWGN channel it is possible to achieve very high reliability (RWER $< 10^{-6}$) with very little parity-check information and resilience techniques in the video codec would hardly be necessary [SFLG00]. For the mobile fading channel, however, the effective use of FEC is limited when restricting the end-to-end delay and the use of error resilience techniques in the source codec is very important.

By increasing the redundancy of the channel code, the available bit-rate for the source coder is reduced. Figure 6.4 shows rate distortion plots obtained from coding experiments with the QCIF sequences *Foreman* as well as *Mother & Daughter*.

Both coders are run with a rate-control enforcing a fixed number of bits per frame when coding 210 frames of video while skipping 2 out of 3 frames. The rate-control employed here adjusts the macroblock quantizer value on the frame-basis so as to hit the target bit-rate. The first 10 frames are excluded from the rate-distortion measurements to avoid the transition phase at the beginning of the sequence since long-term memory MCP with $M = 10$ reference frames is investigated. The two curves in each of the two plots compare

- **TMN-10**: the test model of the H.263 standard with Annexes D, F, I, J, and T enabled.

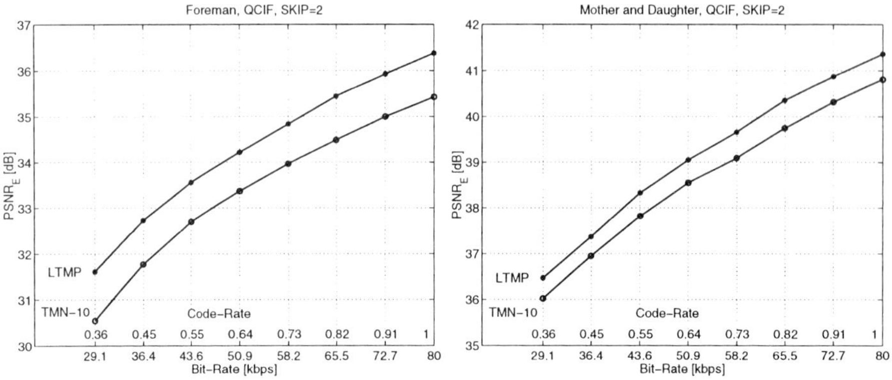

Figure 6.4. Average PSNR vs. average bit-rate or code rate for the sequences *Foreman* (left) and *Mother & Daughter* (right). The two curves relate to the two codecs compared: (i) TMN-10, the test model of the H.263 standard with Annexes D, F, I, J, and T enabled; (ii) long-term memory MCP: the long-term memory prediction coder with 10 frames also utilizing Annexes D, F, I, J, and T.

- **LTMP**: the long-term memory MCP coder with $M = 10$ reference frames also utilizing Annexes D, F, I, J, and T.

The two plots in Fig. 6.4 illustrate various aspects. The PSNR values differ about 5 dB comparing the lowest bit-rate point to the highest bit-rate for both sequences and both codecs. Further, the level of the PSNR values is about 5 dB higher for the sequence *Mother & Daughter* compared to the sequence *Foreman*. This is because, the *Foreman* sequence shows much more motion and high frequency content than the sequence *Mother & Daughter*. The two sequences are chosen as test sequences throughout the chapter because they are considered as extreme cases in the spectrum of low bit-rate video applications.

Finally, the improved coding performance of long-term memory prediction is demonstrated. The bit-rate savings obtained by the long-term memory codec against TMN-10 are 18 % for the sequence *Foreman* when measuring at equal PSNR of 34 dB and 12 % for the sequence *Mother & Daughter* when measuring at 39 dB.

6.3.3 RESULTS WITHOUT FEEDBACK

The first set of simulation results is presented for the case when there is no feedback available. For that, the TMN-10 coder is compared with the long-term memory MCP coder both employing the error modeling approach.

In Fig. 6.5, the average PSNR measured at the encoder (PSNR_E) is depicted versus various code rates for the sequence *Foreman*. The left-hand side plot corresponds to TMN-10 while the right-hand side plot shows results from the long-term memory MCP coder. Both coders are operated under similar condi-

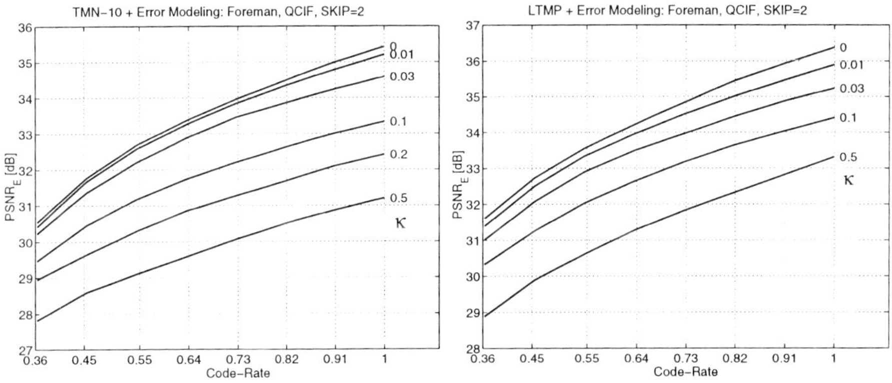

Figure 6.5. Average encoder PSNR vs. code rate for the sequence *Foreman* when running the TMN-10 coder (left) and the long-term memory MCP coder (right). By increasing κ, the error modeling term is amplified resulting in reduced coding performance.

tions as for the results in Fig. 6.4. The various curves correspond to different values of κ, the weight of the transmission error modeling term. The case $\kappa = 0$ is the same as for the curves plotted in Fig. 6.4. Comparing to this case, a significant degradation in terms of coding efficiency can be observed with increasing values of κ for both coders. This performance loss is explained by the additional cost term in (6.13) and (6.14) resulting in increased amounts of INTRA-coded macroblocks and modified motion vectors and, for the long-term memory MCP coder, picture reference parameters.

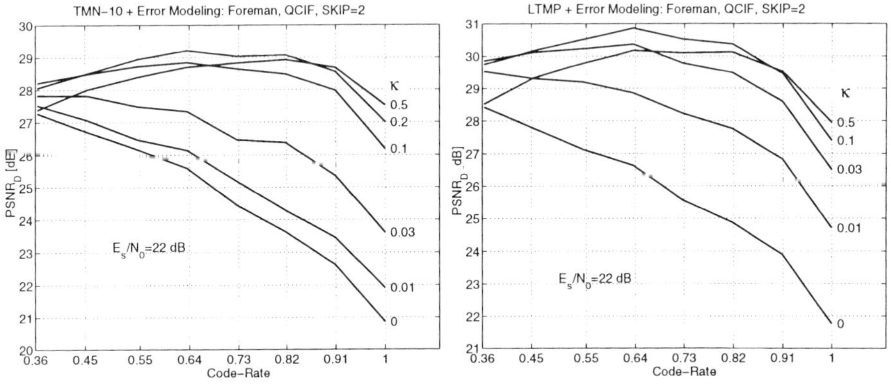

Figure 6.6. Average decoder PSNR vs. code rate for the sequence *Foreman* when running the TMN-10 coder (left) and the long-term memory MCP coder (right).

Figure 6.6 shows the corresponding average PSNR values measured at the decoder (PSNR$_D$). Each bit-stream is transmitted to the decoder via the error-prone channel. This experiment is repeated 30 times using shifted versions of the bit error sequence that corresponds to the fading channel generated with noise-spectral-energy $E_s/N_0 = 22$ dB under the conditions described above. Again, the left-hand side plot shows TMN-10 results while the right-hand side plot depicts results from the long-term memory MCP codec both incorporating the error modeling term using κ. Obviously, the sacrifice at the encoder side pays off at the decoder side. In other words, the weighting factor κ can be used to trade-off coding efficiency and error resilience.

Although the optimum κ generally increases with the code rate and hence with RWER, there is no direct relationship between κ and RWER. To some extent, this results from the fact that κ and RWER describe the loss of probability of different entities, i.e., macroblocks and words. Further, the described simplifications for the estimation of the expected transmission error distortion make it difficult to provide an exact mapping of RWER to κ. Nevertheless, such a mapping is important in practice to operate the codec at the optimal point and is the subject of future research. On the other hand, code rate and κ value can be traded off against each other over a wide range leading to a plateau of similar values of decoder PSNR for various selected pairs of code rate and κ. This feature is especially important when there is no feedback available about the status of a time-varying channel.

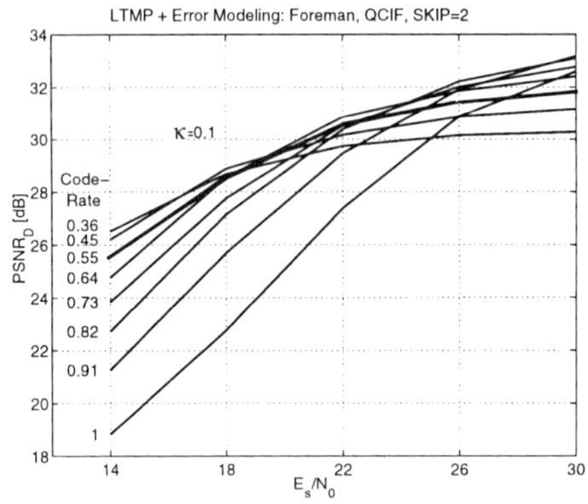

Figure 6.7. Average decoder PSNR vs. E_s/N_0 for the sequence *Foreman* when running the long-term memory MCP coder for a fixed error weighting but various code rates.

The trade-off between code rate and the error modeling term for various channel conditions is illustrated in Fig. 6.7. Average decoder PSNR is shown versus various levels of channel interference expressed by E_s/N_0. The plot is obtained by running the long-term memory MCP codec with a fixed value of $\kappa = 0.1$. The various curves relate to 8 code rates that are equidistantly spaced in the range $32/88 \ldots 1$. Obviously, avoiding channel coding entirely is not advantageous if there are single bit errors as is the case in the simulations. The optimum coding redundancy, of course, depends on the quality of the channel. Nevertheless, the curves corresponding to medium code rates are close to the maximum of the achievable decoder PSNR values indicating that the choice of the code rate is not that critical when combined with the error modeling approach. For example, a code rate of 0.55 provides reasonable performance for the whole range of E_s/N_0 as illustrated by the bold curve in Fig. 6.7.

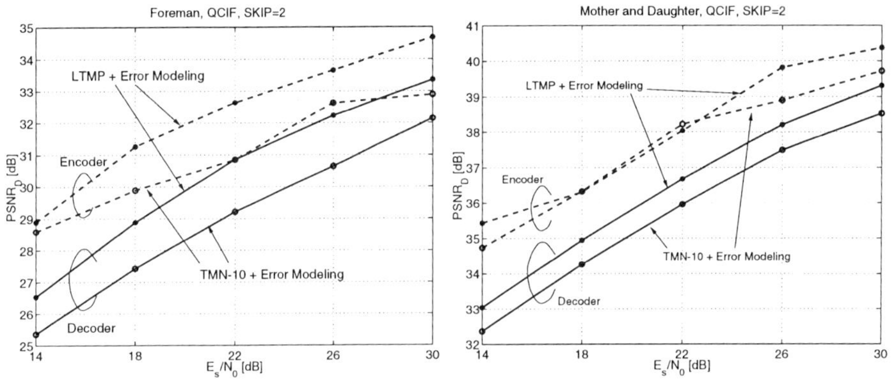

Figure 6.8. Average decoder PSNR vs. E_s/N_0 for the sequences *Foreman* (left) and *Mother & Daughter* (right) for the optimal code rate and error model parameter κ.

In Fig. 6.8, the best performance in terms of maximum decoder PSNR achievable is compared when varying over code rate as well as κ. For that, several simulations have been conducted to sample the parameter space. More precisely, the code rate is varied over 8 values that are equidistantly spaced in the range $32/88 \ldots 1$. The error modeling weight κ is varied over values $0, 0.01, 0.02, 0.03, 0.04, 0.05, 0.10, 0.20, 0.30, 0.40, 0.50$. For each of these 88 pairs of code rate and κ, a bit-stream is encoded using the TMN-10 as well as the long-term memory MCP coder. Each bit-stream is transmitted to the decoder via the error-prone channel. This experiment is repeated 30 times for each channel using shifted versions of the bit error sequences that correspond to E_s/N_0 values of 14, 18, 22, 26, and 30 dB. The dashed lines show encoder PSNR that corresponds to the maximum average PSNR measured at the decoder which is depicted with solid lines. Evaluating decoder PSNR, the long-term memory MCP coder outperforms the TMN-10 coder for all channel conditions. For example, the PSNR gain obtained for the sequence *Foreman* is 1.8 dB at $E_s/N_0 = 22$ dB. Correspondingly, a saving of 4 dB in terms of power efficiency is obtained.

Finally, the validity of the error modeling term in (6.12) as an approximation of the expected transmission error distortion is investigated. As mentioned before, only a subset of the entire error event tree in Fig. 6.1 is considered in (6.12). Hence, only an *approximate error modeling* value of the expected transmission error distortion is used resulting in suboptimal performance. A more accurate error modeling term for the presented simulation scenario is given by the average divergence between encoder and decoder for those 30 simulations for which the performance evaluations are conducted. Note that this *ensemble-based error modeling* term provides a too optimistic anchor, since this scenario can not be realized in practice because those realizations are not known. Nevertheless, within the simulation framework in this chapter, it provides a baseline for comparison.

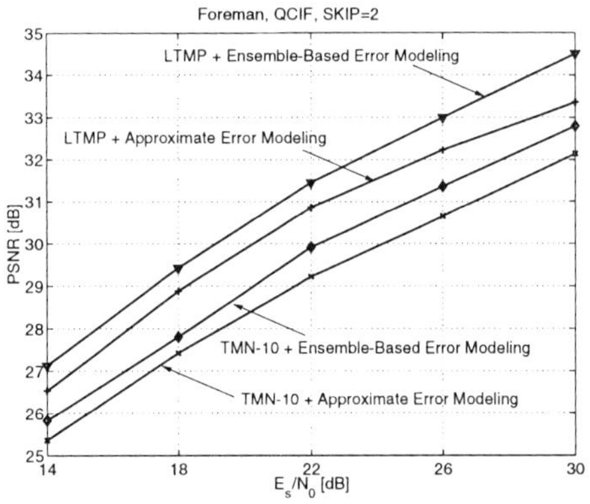

Figure 6.9. Average decoder PSNR vs. E_s/N_0 for the sequence *Foreman* when comparing the results of optimal code rate and κ for approximate error modeling to ensemble-based error modeling.

Figure 6.9 compares the results for the *approximate error modeling* to the *ensemble-based error modeling*. The curves for the *approximate error modeling* are identical to those in Fig. 6.8. In the experiments, the *ensemble-based error modeling* is conducted by locally decoding the received bit-streams that are transmitted over the 30 channel realizations at the encoder. Having the 30 decoding results available at the encoder, the squared difference to the correctly decoded picture is computed and averaged over all 30 cases. This squared difference is employed as D_{ERR} with $\kappa = 1$ in the optimization criteria for the motion estimation (6.13) and mode decision (6.14).

The *ensemble-based error modeling* provides improved performance in terms of decoder PSNR. The main focus of this chapter, however, is not the error modeling scheme. Rather, the error resilience characteristics of long-term memory prediction are investigated in contrast to single-frame prediction. The maximum gains for the *ensemble-based error modeling* in comparison to the *approximate error modeling* are less than 1 dB for large E_s/N_0 values indicating the validity of the approximation. Nevertheless, the impact of the accuracy for the transmission error modeling on rate-distortion performance are subject to future research.

6.3.4 EXPERIMENTAL RESULTS WITH FEEDBACK

In the following set of experiments, a feedback channel is utilized. Such a feedback channel indicates which parts of the bit-stream were received intact and/or which parts of the video signal could not be decoded and had to be concealed. The decoder sends a NACK for an erroneously received GOB and an ACK for a correctly received GOB. For the simulations, it is assumed that the feedback channel is error-free. The round trip delay is set to be approximately 250 ms, such that feedback is received 3 frames after their encoding.

In Fig. 6.10, three feedback handling strategies are compared for the sequence *Foreman*. The simulation conditions are similar to the previous results in this chapter. The left hand side plot shows results obtained with the TMN-10 coder while the right hand side plot shows results from the long-term memory MCP coder. Note that the feedback handling depends on the video codec used.

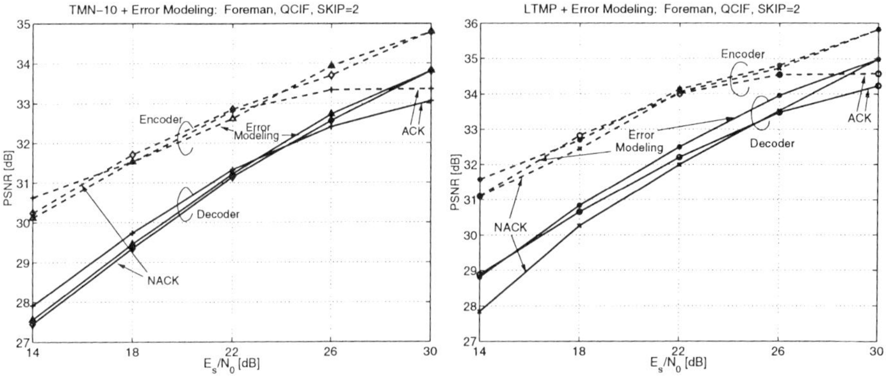

Figure 6.10. Average PSNR vs. E_s/N_0 for the sequence *Foreman* running the TMN-10 coder (left) and the long-term memory MCP coder (right) when feedback is utilized. The various curves correspond to decoder and encoder PSNR for three different feedback-handling strategies.

For the TMN-10 codec, the three feedback-handling schemes are realized as

- **ACK mode**: MCP is conducted without considering the error modeling term ($\kappa = 0$ in (6.13) and (6.14)) by referencing the most recent image for which feedback is available.

- **NACK mode**: MCP is conducted without considering the error modeling term ($\kappa = 0$ in (6.13) and (6.14)) by referencing the most recently decoded frame. Only in case of an error indication via feedback, the image is referenced for which that feedback is received after error concealment.

- **Error Modeling**: As the NACK mode, but motion estimation and mode decision are modified by the transmission error modeling term via setting $\kappa > 0$ in (6.13) and (6.14). The transmission error modeling term D_{ERR} is updated given the feedback messages. For that, the error event tree always starts at the frame for which a feedback message is received and error modeling values are updated for the depending frames using the approximate error modeling approach (see Section 6.2.2).

For the results shown, the parameter space of code rates and κ values is sampled so as to obtain maximum decoder PSNR for the various channel conditions that are given by varying E_s/N_0 over the values $14, 18, 22, 26$, and 30 dB. Evaluating decoder PSNR in the left-hand side plot of Fig. 6.10, the differences between the three schemes are rather small. At $E_s/N_0 = 30$ dB, the NACK mode and the error modeling work best, while at $E_s/N_0 = 26$ dB the error modeling is slightly better. For lower E_s/N_0 values, the ACK mode outperforms the other two schemes.

For the long-term memory MCP codec, the three feedback-handling strategies are implemented as follows.

- **ACK mode**: long-term memory MCP is conducted without considering the error modeling term ($\kappa = 0$ in (6.13) and (6.14)) by referencing the $M = 10$ most decoded recent images for which feedback is available.

- **NACK mode**: long-term memory MCP is conducted without considering the error modeling term ($\kappa = 0$ in (6.13) and (6.14)) by referencing the most recent $M = 10$ decoded frames regardless whether or not feedback is available for them. When an error is indicated via feedback from the decoder, the depending frames are decoded again after error concealment in the feedback frame.

- **Error Modeling**: As the NACK mode, but motion estimation and mode decision are modified by the error modeling term via setting $\kappa > 0$ in (6.13) and (6.14). In addition to concealing the feedback frame and re-decoding of the depending frames, also the error modeling term D_{ERR} is updated for those frames. For that, D_{ERR} is set to zero for the feedback frame because its decoded version at the decoder is known at the encoder. Then, the error event tree starts at the feedback frame and the transmission error estimate D_{ERR} can be updated.

The remaining simulation conditions regarding κ and code rates are the same as for the TMN-10 codec. Here, the differences in terms of decoder PSNR are more visible distinguishing the three concepts. The error modeling approach is superior or achieves similar performance comparing it to the ACK or NACK mode. This is because the ACK or NACK mode in the long-term memory

MCP codec are special cases of the error modeling approach. The ACK mode is incorporated via large values of κ. For large values of κ, reference frames for which no feedback is available are completely avoided since for reference frames with feedback, the term D_{ERR} in (6.13) and (6.14) is set to 0. Hence, ACK mode and error modeling perform equally well for $E_S/N_0 = 14$ dB. On the other hand, the NACK mode is incorporated by simply setting $\kappa = 0$. In this case, both schemes perform equally well for $E_S/N_0 = 30$ dB. Hence, it is not surprising that in the case when ACK and NACK mode perform similarly "well" or rather "poorly," the error modeling approach provides the largest benefit. This can be seen for the results obtained at $E_s/N_0 = 26$ dB.

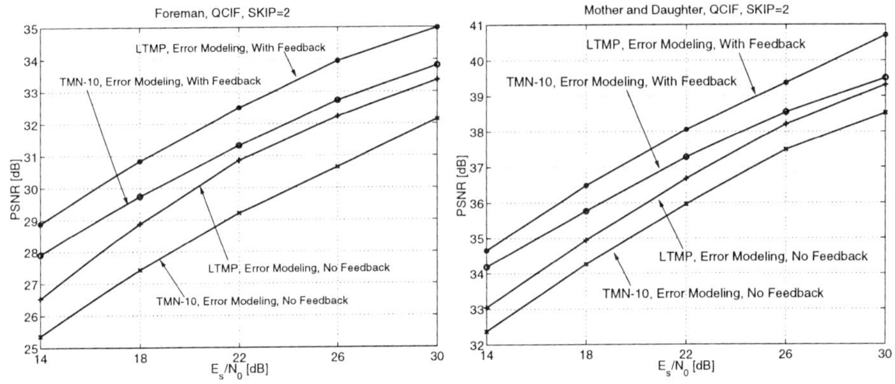

Figure 6.11. Average decoder PSNR vs. E_s/N_0 for the sequences *Foreman* (left) and *Mother & Daughter* (right) for the optimal feedback-handling strategy and without feedback.

Finally, in Fig. 6.11, the gains achievable with the long-term memory MCP codec over the TMN-10 codec are depicted for the feedback case. The results for the case without feedback are shown as well in order to illustrate the error mitigation by feedback. Figure 6.11 depicts these comparisons for the sequences *Foreman* (left) and *Mother & Daughter* (right). The decoder PSNR curves related to the case without feedback are the same as in Fig. 6.8. For the feedback case, the optimum performance points in terms of decoder PSNR are taken from Fig. 6.10 for the *Foreman* sequence. The points for the sequence *Mother & Daughter* are generated in a similar manner.

For the *Foreman* sequence, the error mitigation by feedback in terms of average decoder PSNR is between 1.8 dB at $E_s/N_0 = 30$ dB and 2.5 dB at $E_s/N_0 = 14$ dB. In the feedback case, the long-term memory MCP coder provides a PSNR gain of 1.2 dB compared to the TMN-10 coder. This decoder PSNR gain corresponds to a saving in terms of power efficiency between 3.8 dB at $E_s/N_0 = 30$ dB and 2.5 dB at $E_s/N_0 = 14$ dB. The long-term memory MCP coder without feedback performs close to the TMN-10 coder with feedback for $E_s/N_0 \geq 22$ dB.

6.4 DISCUSSION AND OUTLOOK

The error modeling approach in combination with long-term memory MCP enables an efficient trade-off between rate-distortion performance and error resilience by adjusting one single parameter. This is illustrated by the results that are obtained with and without feedback where the scheme consistently provides improvements in overall transmission performance. But this scheme might be beneficial also for other scenarios. For example, when considering a multi-point transmission with three or four terminals and each terminal can broadcast a feedback about correctly decoded picture content to the remaining terminals. The ACK and NACK mode would have to make hard decisions which reference frames to use. In contrast, the error modeling approach can provide a weighted estimate of the average divergence between the decoded signal at the transmitter and the receivers and can then use this estimate for coder control.

The error modeling utilized in this chapter is an approximation of the expected transmission error distortion. Although results are presented that relate the proposed scheme to a more accurate error modeling approach, open questions remain about the required complexity and the parameter specifications. Note that there has been a proposal for a recursive algorithm to model the expected divergence between encoder and decoder [ZRR00]. The recursive algorithm in [ZRR00] is accurate and has low computational complexity if there is no spatial filtering applied in the video codec. However, if there is spatial filtering as for half-pixel accurate motion compensation [ITU98a], overlapped block motion compensation (Annex F of H.263) or de-blocking filtering (Annex J of H.263) the algorithm has to be extended which may result in significantly increased complexity.

6.5 CHAPTER SUMMARY

In this chapter, long-term memory MCP is proposed for efficient transmission of coded video over noisy transmission channels. The gains of long-term memory MCP are investigated for channels that show random burst errors. A novel approach to coder control is proposed incorporating an estimate of the average divergence between coder and decoder given the statistics of the random channel and the inter-frame error propagation. When employing long-term memory MCP, inter-frame error propagation is considered in the multi-frame buffer which is controlled by the picture reference parameter. Hence, the estimate of the average divergence between coder and decoder is incorporated into the selection of the macroblock modes and the motion vectors including the picture reference parameter.

Experimental results with a Rayleigh fading channel show that long-term memory MCP significantly outperforms the single-frame MCP of the H.263-

based anchor in the presence of error-prone channels for transmission scenarios with and without feedback. In the latter case a PSNR gain at the decoder obtained for the sequence *Foreman* is reported to be up to 1.8 dB. The comparisons are made when fixing the overall bit-rate to 80 kbit/s for both codecs. When a feedback channel is available, the decoder can inform the encoder about successful or unsuccessful transmission events by sending ACKs or NACKs. Upon receipt of feedback, various strategies are known in literature including Error Tracking, ACK and NACK mode. The new coder control unifies these concepts and achieves a trade-off between them by adjusting a simple parameter. Hence, it is not surprising that in the case when ACK and NACK mode perform similarly "well" or rather "poorly," the novel coder control provides the largest benefit. The PSNR gain by the long-term memory scheme compared to single-frame prediction is up to 1.2 dB.

Chapter 7

CONCLUSIONS

The combination of multi-frame MCP with Lagrangian bit-allocation significantly improves the rate-distortion performance of hybrid video coding. For multi-frame prediction, motion compensation is extended from referencing the prior decoded frame to several frames. For that, the motion vector utilized in block-based motion compensation is extended by a picture reference parameter. The picture reference parameter is transmitted as side information, requiring additional bit-rate. The Lagrangian bit-allocation controls the trade-off between the bit-rate for the picture reference parameter and the bit-rate for the MCP residual. Thereby, the additional reference pictures are only utilized if the overall rate-distortion performance is improved.

The Lagrangian bit-allocation employs rate-constrained motion estimation and coding mode decision, which are combined into an efficient control scheme for a video coder as shown for ITU-T Recommendation H.263. Moreover, a new approach for choosing the coder control parameter is presented and its efficiency is demonstrated. The Lagrangian bit-allocation that is developed in this book led to the creation of a new encoder recommendation for ITU-T Recommendation H.263+ which is called TMN-10. The comparison to TMN-9, the predecessor of TMN-10, shows that a bit-rate reduction up to 10 % can be achieved. Moreover, the new test model of H.26L basically follows the presented approach.

Long-term memory MCP is an efficient method to exploit long-term statistical dependencies in video sequences. For long-term memory MCP, multiple past decoded pictures are referenced for motion compensation. A statistical model for the prediction gain provides the insight that the PSNR improvements in dB are roughly proportional to the log-log of the number of reference frames. The integration of long-term memory MCP into an H.263-based hybrid video codec shows that significant improvements in rate-distortion efficiency can be obtained. For that, the Lagrangian bit-allocation scheme is extended to long-term memory MCP. When considering 34 dB reproduction quality and employing 10 reference frames, an average bit-rate reduction of 12 % against TMN-10 can be observed for the set of test sequences. This represents a large variety of video content ranging from low-motion sequences with a fixed camera to sequences with a large amount of motion, including a moving camera position and focal length change. When employing 50 reference frames, the bit-rate savings against TMN-10 are around 17 %, the minimal bit-rate savings inside the test set are 13 %, while the maximal bit-rate savings are reported to be up to 23 %. These bit-rate savings relate to PSNR gains between 0.7 to 1.8 dB. For some image sequences, very significant bit-rate savings of more than 60 % can be achieved. Based on the techniques and results presented, the long-term memory MCP scheme has been accepted as Annex U of ITU-T Recommendation H.263, version 3. The currently ongoing project of the ITU-T Video Coding Experts Group, H.26L, incorporates long-term memory MCP from the very beginning as an integral part.

The concept of long-term memory MCP can be taken further by extending the multi-frame buffer with warped versions of decoded frames. Affine motion parameters describe the warping. The idea of reference picture warping can be regarded as an alternative approach to assigning affine motion parameters to large image segments with the aim of a rate-distortion efficient motion representation. For that, the Lagrangian coder control adapts the number of affine motion parameter sets to the input statistics. Experimental results validate the effectiveness of the new approach. When warping the prior decoded frame, average bit-rate savings of 15 % against TMN-10 are reported for the case that 20 additional reference pictures are warped. Within the set of test sequences, the bit-rate savings vary from 6 to 25 %. For the measurements, reconstruction PSNR is identical to 34 dB for all cases considered. Further improvements can be obtained when in addition to warping the prior decoded frame also older decoded frames are warped. This combination of long-term memory MCP and reference picture warping provides almost additive rate-distortion gains. When employing 10 decoded reference frames and 20 warped reference pictures, average bit-rate savings of 24 % can be obtained. The minimal bit-rate savings inside the test set are 15 %, while the maximal bit-rate savings are up to 35 %. These bit-rate savings correspond to gains in PSNR between 0.8 and 3 dB. For some cases, the combination of affine and long-term memory MCP provides more than additive gains.

The novel techniques for fast multi-frame motion estimation show that the computational requirements can be reduced by more than an order of magnitude, while maintaining all or most of the improvements in coding efficiency. The main idea investigated is to pre-compute data about the search space that can be used to either avoid considering certain positions or to reduce the complexity for evaluating distortion. When a picture is decoded and included into the set of M reference pictures, the fact can be exploited that $M - 1$ of the pictures and the associated pre-computed data remain unchanged in the multi-frame buffer. Another important fact that is exploited is that many blocks in the multi-frame buffer are quite similar and can thus be excluded from the search space. Experimental results show that in comparison to full-search motion estimation, significant reductions in computation time are achieved, which indicate that the increased computational complexity for multi-frame motion estimation is not an obstacle to practical systems.

The efficiency of long-term memory MCP is investigated for channels that show random burst errors. A novel approach to coder control is proposed incorporating an estimate of the average divergence between coder and decoder given the statistics of the random channel and the inter-frame error propagation. When employing long-term memory MCP, inter-frame error propagation is considered in the multi-frame buffer, which is controlled by the picture reference parameter. Hence, the estimate of the average divergence between coder and decoder is incorporated into the selection of the macroblock modes and the motion vectors including the picture reference parameter. The experimental results with a Rayleigh fading channel show that long-term memory MCP significantly outperforms the single-frame MCP of the H.263-based anchor in the presence of error-prone channels for transmission scenarios with and without feedback. In the latter case a PSNR gain at the decoder obtained for the sequence *Foreman* is reported to be up to 1.8 dB. The comparisons are made when fixing the overall transmission bit-rate to 80 kbit/s for both codecs. When a feedback channel is available, the decoder can inform the encoder about successful or unsuccessful transmission events by sending ACKs or NACKs. Various feedback-handling strategies are known in literature including Error Tracking, ACK and NACK mode. The new coder control unifies these concepts and achieves a trade-off between them by adjusting a simple parameter. The PSNR gain by the long-term memory scheme compared to single-frame prediction is up to 1.2 dB.

The results of this book indicate that multi-frame prediction can significantly enhance the rate-distortion efficiency of video transmission systems. Whether the developed approaches should be included into a practical video transmission system, and if so, which parameter settings should be used, has to be judged considering the available resources. The results in this book are presented to give bounds and guidance about the benefits and drawbacks of the new techniques. With the new flexible syntax, some improvements with multi-frame MCP over single-frame MCP can be obtained at a very little overhead in complexity. The costs due to the increase in complexity for the other cases become increasingly smaller by advances in semiconductor technology. It is to be hoped that the contributions of this book encourage the multi-frame MCP approach to be considered as an integral part in the design of future video transmission systems.

Appendix A
Simulation Conditions

A.1 DISTORTION MEASURES

The performance evaluation of video transmission systems and bit allocation in the video encoder requires an ability to measure distortion. However, the distortion that a human observer perceives in coded visual content is a very difficult quantity to measure, as the characteristics of the human visual system are complex and not well understood. This problem is aggravated in video coding, because the addition of the temporal domain relative to still-picture coding further complicates the issue. In practice, simple objective distortion measures such as SSD or its equivalents known as MSE or PSNR are used in most actual comparisons. They are defined by

$$\mathrm{SSD} = \sum_{(x,y)\in\boldsymbol{\mathcal{A}}} |s[x,y,t] - \acute{s}[x,y,t]|^2 \tag{A.1}$$

$$\mathrm{MSE} = \frac{1}{|\boldsymbol{\mathcal{A}}|}\mathrm{SSD} \tag{A.2}$$

$$\mathrm{PSNR} = 10\log_{10}\frac{255^2}{\mathrm{MSE}}\ \mathrm{dB} \tag{A.3}$$

Another distortion measure in common use (since it is often easier to compute) is the SAD

$$\mathrm{SAD} = \sum_{(x,y)\in\boldsymbol{\mathcal{A}}} |s[x,y,t] - \acute{s}[x,y,t]| \tag{A.4}$$

where s and $\acute{s}$ are the luminance signal of the actual and approximated pictures, the set $\boldsymbol{\mathcal{A}}$ contains the set of pixel locations over which distortion is measured, and $|\boldsymbol{\mathcal{A}}|$ specifies the number of pixels in $\boldsymbol{\mathcal{A}}$. The distortion measures in (A.1)-(A.4) are used throughout this book.

In case the received bit-stream contains random errors, expected distortion measures are utilized to give a measure of system performance. In practice, this will be realized via transmitting the video signal a sufficient number of times over the error prone channel and then averaging the PSNR values for the various trials.

A.2 TEST SEQUENCES

The experiments in this book are conducted using the QCIF test sequences and conditions in Tab. A.1. The sequences and test conditions are almost identical to those that are maintained by the ITU-T Video Coding Experts Group. This set of sequences has been chosen so as to represent a wide variety of statistical dependencies and different types of motion and texture.

Sequence Name	Abbreviation	Number of Frames	Frame Skip	Global Motion
Foreman	fm	400	2	Yes
Mobile & Calendar	mc	300	2	Yes
Stefan	st	300	2	Yes
Tempete	te	260	1	Yes
Container Ship	cs	300	2	No
Mother & Daughter	md	300	2	No
News	nw	300	2	No
Silent Voice	si	300	1	No

Table A.1. Test sequences and simulation conditions.

The first four sequences contain a large amount of motion including a moving camera position and focal length change. The last four sequences are low motion sequences with a fixed camera. This set was chosen so as to cover a broad range of possible scenes that might occur in applications such as video conferencing or video streaming.

In all experiments, bit-streams are generated that are decodable producing the same PSNR values at encoder and decoder. The first frame of the image sequence is coded in INTRA mode followed by INTER-coded pictures. In INTER pictures the macroblocks can either be coded predictively using one of the INTER macroblock modes or as INTRA blocks. In the simulations, the first intra-coded frame is identical for all cases considered.

Appendix B
Computation of Expected Values

Note:This appendix results from joint work with Joachim Eggers.

Our objective is to compute mean and variance of $\mathcal{Y}_{1,2}$. The random variable $\mathcal{Y}_{1,2}$ is assigned to a random experiment involving the random variable $\mathcal{X}^2$ that is denoted as

$$\mathcal{X}^2 = (\mathcal{X}_1, \mathcal{X}_2)^T. \tag{B.1}$$

The random experiment consists of a minimization that is conducted by drawing an 2-tuple $\mathcal{X}^2 = (\mathcal{X}_1, \mathcal{X}_2)$ and choosing the minimum element. This minimum element is considered as the outcome of another random experiment and the associated random variable is called $\mathcal{Y}_{1,2}$.

The expected value of $\mathcal{Y}_{1,2}$ can be expressed via conditional expectations

$$E\{\mathcal{Y}_{1,2}\} = \int_{-\infty}^{\infty} E\{\mathcal{Y}_{1,2}|\mathcal{X}_1\} f_{\mathcal{X}_1}(x_1)\mathrm{d}x_1. \tag{B.2}$$

The minimization can be incorporated into the computation of a conditional expected value $E\{\mathcal{Y}_{1,2}|\mathcal{X}_1\}$ as illustrated in Fig. B.1.

For a given value x_1, the conditional expected value $E\{\mathcal{Y}_{1,2}|\mathcal{X}_1\}$ is split into two cases:

- $x_2 \geq x_1$: i.e., the minimization result is x_1. The integral over the shaded part in Fig. B.1 corresponds to the probability of this outcome of the experiment.
- $x_2 < x_1$: i.e., the minimization result is x_2. The integral over the remaining (non-shaded) part of the PDF gives the probability of this case.

Hence, the conditional expected value is given as

$$E\{\mathcal{Y}_{1,2}|\mathcal{X}_1\} = \int_{-\infty}^{x_1} x_2 f_{\mathcal{X}_2|\mathcal{X}_1}(x_1, x_2)\mathrm{d}x_2 + x_1 \int_{x_1}^{\infty} f_{\mathcal{X}_2|\mathcal{X}_1}(x_1, x_2)\mathrm{d}x_2. \tag{B.3}$$

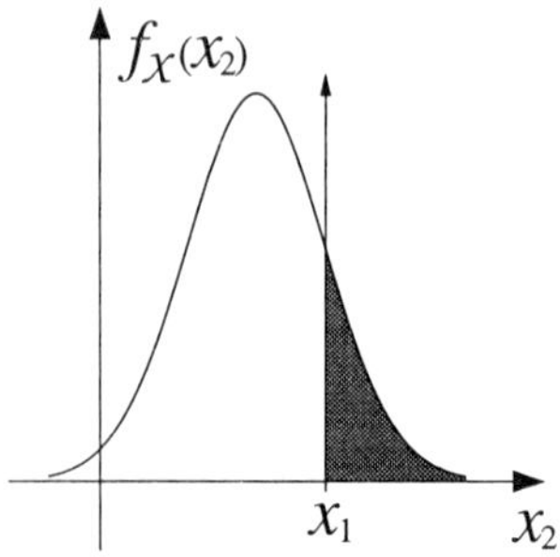

Figure B.1. Conditional expected value.

By plugging (B.3) into (B.2) and some simple manipulations we arrive at

$$E\{\mathcal{Y}_{1,2}\} = E\{\mathcal{X}_1\} - \int_{-\infty}^{\infty} g(x_1) f_{\mathcal{X}_1}(x_1) \mathrm{d}x_1, \tag{B.4}$$

with

$$g(x_1) = \int_{-\infty}^{x_1} (x_1 - x_2) f_{\mathcal{X}_2|\mathcal{X}_1}(x_1, x_2) \mathrm{d}x_2. \tag{B.5}$$

The integrals can be solved analytically in case of jointly normal distributions yielding the mean difference $\Delta_{1,2}$ via minimization as

$$\Delta_{1,2} = E\{\mathcal{X}_1\} - E\{\mathcal{Y}_{1,2}\} = \mu - E\{\mathcal{Y}_{1,2}\} = \sigma\sqrt{\frac{1-\rho}{\pi}}. \tag{B.6}$$

Employing similar arguments, we obtain an expression for the variance ratio which reads

$$\xi_{1,2}^2 = \frac{E\left\{\mathcal{Y}_{1,2}^2\right\} - E\{\mathcal{Y}_{1,2}\}^2}{E\{\mathcal{X}_1^2\} - E\{\mathcal{X}_1\}^2} = \frac{E\left\{\mathcal{Y}_{1,2}^2\right\} - E\{\mathcal{Y}_{1,2}\}^2}{\sigma^2} = 1 - \frac{1-\rho}{\pi}. \tag{B.7}$$

References

[BG96] M. Budagavi and J. D. Gibson. Multiframe Block Motion Compensated Video Coding for Wireless Channels. In *Proceedings of the Asilomar Conference on Signals, Systems, and Computers*, volume 2, pages 953–957, Asilomar, CA, USA, November 1996.

[BG97] M. Budagavi and J. D. Gibson. Error Propagation in Motion Compensated Video over Wireless Channels. In *Proceedings of the IEEE International Conference on Image Processing*, volume 2, pages 89–92, Santa Barbara, CA, USA, October 1997.

[BG98] M. Budagavi and J. D. Gibson. Random Lag Selection in Multiframe Motion Compensation. In *Proceedings of the IEEE International Symposium on Information Theory*, Boston, MA, USA, August 1998.

[BGM+98] E. Berruto, M. Gudmundson, R. Menolascino, W. Mohr, and M. Pizarroso. Research Activities on UMTS Radio Interface, Network Architectures, and Planning. *IEEE Communications Magazine*, 36(2):82–95, February 1998.

[CAS+96] C. K. Cheong, K. Aizawa, T. Saito, M. Kaneko, and H. Harashima. Structural Motion Segmentation for Compact Image Sequence Representation. In *Proceedings of the SPIE Conference on Visual Communications and Image Processing*, volume 2727, pages 1152–1163, Orlando, FL, USA, March 1996.

[CEGK98] G. Cote, B. Erol, M. Gallant, and F. Kossentini. H.263+: Vide Coding at Low Bit Rates. *IEEE Transactions on Circuits and Systems for Video Technology*, 8(7):849–866, November 1998.

[Che95] C. Chen. Error Detection and Concealment with an Unsupervised MPEG2 Video Decoder. *Journal of Visual Communication and Image Representation*, 6(3):265–278, September 1995.

[Che99] W. Y. Chen. The Development and Standardization of Asymmetrical Digital Subscriber Line. *IEEE Communications Magazine*, 37(5):68–72, May 1999.

[CKS96] W. C. Chung, F. Kossentini, and M. J. T. Smith. An Efficient Motion Estimation Technique Based on a Rate-Distortion Criterion. In *Proceedings of the IEEE International Conference on Acoustics, Speech and Signal Processing*, pages 1926–1929, Atlanta, GA, USA, May 1996.

[CLG89] P. A. Chou, T. Lookabaugh, and R. M. Gray. Entropy-Constrained Vector Quantization. *IEEE Transactions on Acoustics, Speech and Signal Processing*, 37(1):31–42, January 1989.

[CW96] M. C. Chen and A. N. Willson. Rate-Distortion Optimal Motion Estimation Algorithm for Video Coding. In *Proceedings of the IEEE International Conference on Acoustics, Speech and Signal Processing*, volume 4, pages 2096–2099, Atlanta, GA, USA, May 1996.

[CW98] M. C. Chen and A. N. Willson. Rate-Distortion Optimal Motion Estimation Algorithms for Motion-Compensated Transform Video Coding. *IEEE Transactions on Circuits and Systems for Video Technology*, 8(2):147–158, April 1998.

[Die91] N. Diehl. Object-Oriented Motion Estimation and Segmentation in Image Sequences. *Signal Processing: Image Communication*, 3(1):23–56, January 1991.

[DM96] F. Dufaux and F. Moscheni. Background Mosaicking for Low Bit Rate Video Coding. In *Proceedings of the IEEE International Conference on Image Processing*, volume 3, pages 673–676, Lausanne, Switzerland, September 1996.

[DS95] J.-L. Dugelay and H. Sanson. Differential Methods for the Identification of 2D and 3D Motion Models in Image Sequences. *Signal Processing: Image Communication*, 7(1):105–127, March 1995.

[Eve63] H. Everett III. Generalized Lagrange Multiplier Method for Solving Problems of Optimum Allocation of Resources. *Operations Research*, 11:399–417, 1963.

[EWG00] P. Eisert, T. Wiegand, and B. Girod. Model-Aided Coding: A New Approach to Incorporate Facial Animation into Motion-Compensated Video Coding. *IEEE Transactions on Circuits and Systems for Video Technology*, 10(1):344–358, April 2000.

[FGV98] N. Färber, B. Girod, and J. Villasenor. Extensions of the ITU-T Recommendation H.324 for Error-Resilient Video Transmission. *IEEE Communications Magazine*, 36(6):120–128, June 1998.

[FNI96] S. Fukunaga, T. Nakai, and H. Inoue. Error-Resilient Video Coding by Dynamic Replacing of Reference Pictures. In *GLOBECOM'96*, volume 3, pages 1503–1508, November 1996.

[FSG96] N. Färber, E. Steinbach, and B. Girod. Robust H.263 Compatible Video Transmission Over Wireless Channels. In *Proceedings of the Picture Coding Symposium*, pages 575–578, 1996.

[FSG99] N. Färber, K. W. Stuhlmüller, and B. Girod. Analysis of Error Propagation in Hybrid Video Coding with Application to Error Resilience. In *Proceedings of the IEEE International Conference on Image Processing*, volume 2, pages 550–554, Kobe, Japan, October 1999.

[FVC87] E. Francois, J.-F. Vial, and B. Chupeau. Coding Algorithm with Region-Based Motion Compensation. *IEEE Transactions on Circuits and Systems for Video Technology*, 7(1):97–108, February 1987.

[FWG98] M. Flierl, T. Wiegand, and B. Girod. A Local Optimal Design Algorithm for Block-Based Multi-Hypothesis Motion-Compensated Prediction. In *Proceedings of the IEEE Data Compression Conference*, pages 239–248, Snowbird, USA, March 1998.

[FWG00a] M. Flierl, T. Wiegand, and B. Girod. A Video Codec Incorporating Block-Based Multi-Hypothesis Motion-Compensated Prediction. In *Proceedings of the SPIE Conference on Visual and Image Processing*, Perth, Australia, June 2000.

[FWG00b] M. Flierl, T. Wiegand, and B. Girod. Rate-Constrained Multi-Hypothesis Motion-Compensated Prediction for Video Coding. In *Proceedings of the IEEE International Conference on Processing*, Vancouver, Canada, October 2000.

[GF99] B. Girod and N. Färber. Feedback-Based Error Control for Mobile Video Transmission. *Proceedings of the IEEE*, 97(10):1707–1723, October 1999.

[Gir87] B. Girod. The Efficiency of Motion-Compensating Prediction for Hybrid Coding of Video Sequences. *IEEE Journal on Selected Areas in Communications*, 5(7):1140–1154, August 1987.

[Gir93] B. Girod. Motion-Compensating Prediction with Fractional-Pel Accuracy. *IEEE Transactions on Communications*, 41(4):604–612, April 1993.

[Gir94] B. Girod. Rate-Constrained Motion Estimation. In *Proceedings of the SPIE Conference on Visual Communications and Image Processing*, volume 2308, pages 1026–1034, Chicago, IL, USA, September 1994.

[Gir00] B. Girod. Efficiency Analysis of Multi-Hypothesis Motion-Compensated Prediction. *IEEE Transactions on Image Processing*, 9(2):173–183, February 2000.

[GL94] M. Ghanbari and T. K. B. Lee. Use of 'Late Cells' for ATM Video Enhancement. In *Packet Video Workshop*, pages I2.1–I2.4, Portland, OR, USA, September 1994.

[GP68] H. Gish and J. N. Pierce. Asymptotically Efficient Quantizing. *IEEE Transactions on Information Theory*, 14:676–683, September 1968.

[GSF97] B. Girod, E. Steinbach, and N. Färber. Performance of the H.263 Video Compression Standard. *Journal of VLSI Signal Processing: Systems for Signal, Image, and Video Technology*, 17:101–111, November 1997.

[Hep90] D. Hepper. Efficiency Analysis and Application of Uncovered Background Prediction in a Low Bit Rate Image Coder. *IEEE Transactions on Communications*, 38(9):1578–1584, September 1990.

[HM92] P. Haskell and D. Messerschmitt. Resynchronization of Motion-Compensated Video Affected by ATM Cell Loss. In *Proceedings of the IEEE International Conference on Acoustics, Speech and Signal Processing*, volume 3, pages 545–548, 1992.

[Hor86] B. K. P. Horn. *Robot Vision.* The MIT Press, McGraw-Hill Book Company, USA, 1986.

[Höt89] M. Hötter. Differential Estimation of the Global Motion Parameters Zoom and Pan. *Signal Processing*, 16(3):249–265, March 1989.

[HPL98] R. O. Hinds, T.N. Pappas, and J. S. Lim. Joint Block-Based Video Source/Channel Coding for Packet-Switched Networks. In *Proceedings of the SPIE Conference on Visual Communications and Image*

Processing, volume 3309, pages 124–133, San Jose, CA, USA, January 1998.

[HS81] B. K. P. Horn and B. G. Schunck. Determining Optical Flow. *Artificial Intelligence*, 17(1-3):185–203, 1981.

[HT88] M. Hötter and R. Thoma. Image Segmentation Based on Object Oriented Mapping Parameter Estimation. *Signal Processing: Image Communication*, 15(3):315–334, October 1988.

[HW98] S.-C. Han and J. W. Woods. Adaptive Coding of Moving Objects for Very Low Bit Rates. *IEEE Journal on Selected Areas in Communications*, 16(1):56–70, January 1998.

[ISO93] ISO/IEC JTC1. Coding of Moving Pictures and Associated Audio for Digital Storage Media at Up to About 1.5 Mbit/s — Part 2: Video. International Standard, March 1993.

[ISO94] ISO/IEC JTC1 13818-2; ITU-T Recommendation H.262. Generic Coding of Moving Pictures and Associated Audio Information — Part 2: Video. International Standard, November 1994.

[ISO96a] ISO/IEC JTC1/SC29/WG11 MPEG96/M0654. Core Experiment of Video Coding with Block-Partitioning and Adaptive Selection of Two Frame Memories (STFM / LTFM). December 1996.

[ISO96b] ISO/IEC JTC1/SC29/WG11 MPEG96/M0768, Iterated Systems Inc. An Error Recovery Strategy for Videophone Applications. March 1996.

[ISO97a] ISO/IEC JTC1/SC29/WG11 MPEG96/M1686. Core experiment on global motion compensation. Submitted to Video Subgroup, February 1997.

[ISO97b] ISO/IEC JTC1/SC29/WG11 MPEG96/N1648. Core Experiment on Sprites and GMC. April 1997.

[ISO98a] ISO/IEC JTC1/SC29/WG11 MPEG98/N2202. Committee Draft. March 1998.

[ISO98b] ISO/IEC JTC1/SC29/WG11 N2202. Committee Draft. March 1998.

[ITU] ITU-T Recommendation H.120. Codec for Videoconferencing Using Primary Digital Group Transmission. version 1, 1984; version 2, 1988.

[ITU92] ITU-T and ISO/IEC JTC1. Digital Compression and Coding of Continuous-Tone Still Images. ISO/IEC 10918-1 | ITU-T Recommendation T.81 (JPEG), September 1992.

[ITU93] ITU-T Recommendation H.261. Video Codec for Audiovisual Services at $p \times 64$ kbit/s. March 1993.

[ITU95] ITU-T, SG15/WP15/1, LBC-95-309, National Semiconductor Corporation. Sub-Videos with Retransmission and Intra-Refreshing in Mobile/Wireless Environments. October 1995.

[ITU96a] ITU-T Recommendation H.263. Video Coding for Low Bitrate Communication. June 1996.

[ITU96b] ITU-T, SG15/WP15/1, LBC-95-033, Telenor R&D. An Error Resilience Method Based on Back Channel Signaling and FEC. Also submitted to ISO/IEC JTC1/SC29/WG11 as contribution MPEG96/M0616, January 1996.

[ITU97] ITU-T/SG16/Q15-C-15. Video Codec Test Model Number 9 (TMN-9). Download via anonymous ftp to: standard.pictel.com/video-site/9712_Eib/q15c15.doc, December 1997.

[ITU98a] ITU-T Recommendation H.263 Version 2 (H.263+). Video Coding for Low Bitrate Communication. January 1998.

[ITU98b] ITU-T/SG16/Q15-D-13, T. Wiegand and B. Andrews. An Improved H.263-Codec Using Rate-Distortion Optimization. Download via anonymous ftp to: standard.pictel.com/video-site/9804_Tam/q15d13.doc, April 1998.

[ITU98c] ITU-T/SG16/Q15-D-40, J. Wen, M. Luttrell, and J. Villasenor. Simulation Results on Trellis-Based Adaptive Quantization. Download via anonymous ftp to: standard.pictel.com/video-site/9804_Tam/q15d40.doc, April 1998.

[ITU98d] ITU-T/SG16/Q15-D-65. Video Codec Test Model, Near Term, Version 10 (TMN-10), Draft 1. Download via anonymous ftp to: standard.pictel.com/video-site/9804_Tam/q15d65.doc, April 1998.

[ITU98e] ITU-T/SG16/Q15-E-44, T. Wiegand, N. Färber, B. Girod, and B. Andrews. Long-Term Memory Motion-Compensated Prediction for Surveillance Applications. Download via anonymous ftp to: standard.pictel.com/ video-site/ 9807_Whi/ q15e44.doc, July 1998.

[ITU99a] ITU-T/SG16/Q15-H-09, K. Zhang and J. Kittler. Proposed Amendments for Annex U on Enhanced Reference Picture Selection. Download via anonymous ftp to: standard.pictel.com/video-site/9712_Ber/q15h09.doc, July 1999.

[ITU99b] ITU-T/SG16/Q15-I-44, T. Wiegand. Proposed Draft for Annex U on Enhanced Reference Picture Selection. Download via anonymous ftp to: standard.pictel.com/ video-site/9910_Red/q15i44.doc, October 1999.

[ITU00] ITU-T Recommendation H.263 Version 2 (H.263++). Video Coding for Low Bitrate Communication. November 2000.

[Jak74] W. C. Jakes. *Microwave Mobile Radio Reception.* Wiley, New York, USA, 1974.

[JJ81] J. R. Jain and A. K. Jain. Displacement Measurement and Its Application in Interframe Image Coding. *IEEE Transactions on Communications*, 29(12):1799–1808, December 1981.

[JKS+97] H. Jozawa, K. Kamikura, A. Sagata, H. Kotera, and H. Watanabe. Two-Stage Motion Compensation Using Adaptive Global MC and Local Affine MC. *IEEE Transactions on Circuits and Systems for Video Technology*, 7(1):75–85, February 1997.

[JN94] N. S. Jayant and P. Noll. *Digital Coding of Waveforms.* Prentice-Hall, Englewood Cliffs, NJ, USA, 1994.

[KLSW97] F. Kossentini, Y.-W. Lee, M. J. T. Smith, and R. Ward. Predictive RD-Constrained Motion Estimation for Very Low Bit Rate Video Coding. *IEEE Journal on Selected Areas in Communications*, 15(9):1752–1763, December 1997.

[KNH97] M. Karczewicz, J. Nieweglowski, and P. Haavisto. Video Coding Using Motion Compensation with Polynomial Motion Vector Fields. *Signal Processing: Image Communication*, 10(3):63–91, July 1997.

[LC95] C.-H. Lee and L.-H. Chen. A Fast Search Algorithm for Vector Quantization Using Mean Pyramids of Codewords. *IEEE Transactions on Communications*, 43(2/3/4):604–612, Feb./Mar./Apr. 1995.

[LD94] J. Lee and B. W. Dickinson. Joint Optimization of Frame Type Selection and Bit Allocation for MPEG Video Coders. In *Proceedings of the IEEE International Conference on Image Processing*, volume 2, pages 962–966, Austin, TX, USA, November 1994.

[LF95] H. Li and R. Forchheimer. A Transform Block-Based Motion Compensation Technique. *IEEE Transactions on Communications*, 43(2):1673–1676, February 1995.

[LS95] W. Li and E. Salari. Successive Elimination Algorithm for Motion Estimation. *IEEE Transactions on Image Processing*, 4(1):105–107, January 1995.

[LT97] Y.-C. Lin and S.-C. Tai. Fast Full-Search Block-Matching Algorithm for Motion-Compensated Video Compression. *IEEE Transactions on Communications*, 45(5):527–531, May 1997.

[LT00] Test Model Long-Term. TML (H.26L) Encoder/Decoder, Version 2.0. Download via anonymous ftp to standard.pictel.com/video-site/0005_Osa/q15j08.doc, May 2000.

[LV96] J. Liao and J. Villasenor. Adaptive Intra Update for Video Coding over Noisy Channels. In *Proceedings of the IEEE International Conference on Image Processing*, volume 3, pages 763–766, Lausanne, Switzerland, October 1996.

[MK85] N. Mukawa and H. Kuroda. Uncovered Background Prediction in Interframe Coding. *IEEE Transactions on Communications*, 33(11):1227–1231, November 1985.

[Mou69] F. W. Mounts. A Video Encoding System With Conditional Picture-Element Replenishment. *The Bell System Technical Journal*, 48(7):2545–2554, September 1969.

[MPG85] H. G. Musmann, P. Pirsch, and H.-J. Grallert. Advances in Picture Coding. *Proceedings of the IEEE*, 73(4):523–548, April 1985.

[NO92] S. Nogaki and M. Ohta. An Overlapped Block Motion Compensation for High Quality Motion Picture Coding. In *Proceedings of the IEEE International Symposium on Circuits and Systems*, volume 1, pages 184–187, San Diego, CA, USA, May 1992.

[OR95] A. Ortega and K. Ramchandran. Forward-Adaptive Quantization with Optimal Overhead Cost for Image and Video Coding with Applications to MPEG Video Coders. In *Proceedings of the IS&T/SPIE, Digital Video Compression: Algorithms and Technologies*, volume 2419, pages 129–138, San Jose, CA, USA, February 1995.

[OS94] M. T. Orchard and G. J. Sullivan. Overlapped Block Motion Compensation: An Estimation-Theoretic Approach. *IEEE Transactions on Image Processing*, 3(5):693–699, September 1994.

[PGH95] J. E. Padgett, C. Günther, and T. Hattori. Overview of Wireless Personal Communications. *IEEE Communications Magazine*, 33(1):29–41, January 1995.

[PM93] W. B. Pennebaker and J. L. Mitchell. *JPEG: Still Image Data Compression Standard.* Van Nostrand Reinhold, New York, USA, 1993.

[Rij96] K. Rijkse. H.263: Video Coding for Low-Bit-Rate Communication. *IEEE Communications Magazine*, 34(12):42–45, December 1996.

[ROV94] K. Ramchandran, A. Ortega, and M. Vetterli. Bit Allocation for Dependent Quantization with Applications to Multiresolution and MPEG Video Coders. *IEEE Transactions on Image Processing*, 3(5):533–545, September 1994.

[San91] H. Sanson. Motion Affine Models Identification and Application to Television Image Sequences. In *Proceedings of the SPIE Conference on Visual Communications and Image Processing*, volume 1605, pages 570–581, 1991.

[SB91] G. J. Sullivan and R. L. Baker. Rate-Distortion Optimized Motion Compensation for Video Compression Using Fixed or Variable Size Blocks. In *Proc. GLOBECOM'91*, pages 85–90, Phoenix, AZ, USA, December 1991.

[SFG97] E. Steinbach, N. Färber, and B. Girod. Standard Compatible Extension of H.263 for Robust Video Transmission in Mobile Environments. *IEEE Transactions on Circuits and Systems for Video Technology*, 7(6):872–881, December 1997.

[SFLG00] K. W. Stuhlmüller, N. Färber, M. Link, and B. Girod. Analysis of Video Transmission over Lossy Channels. *IEEE Journal on Selected Areas in Communications*, 18(6), June 2000.

[SG88] Y. Shoham and A. Gersho. Efficient Bit Allocation for an Arbitrary Set of Quantizers. *IEEE Transactions on Acoustics, Speech and Signal Processing*, 36:1445–1453, September 1988.

[SK96] G. M. Schuster and A. K. Katsaggelos. Fast and Efficient Mode and Quantizer Selection in the Rate Distortion Sense for H.263. In *Proceedings of the SPIE Conference on Visual Communications and Image Processing*, volume 2727, pages 784–795, Orlando, FL, USA, March 1996.

[SK97] G. M. Schuster and A. K. Katsaggelos. A Video Compression Scheme with Optimal Bit Allocation Among Segmentation, Motion, and Residual Error. *IEEE Transactions on Image Processing*, 6(11):1487–1502, November 1997.

[Skl97] B. Sklar. Rayleigh Fading Channels in Mobile Digital Communication Systems, Part I: Characterization. *IEEE Communications Magazine*, 35(7):90–100, September 1997.

[SSO99] A. Smolic, T. Sikora, and J.-R. Ohm. Long-Term Global Motion Estimation and Its Application for Sprite Coding, Content Description, and Segmentation. *IEEE Transactions on Circuits and Systems for Video Technology*, 9(8):1227–1242, December 1999.

[Sul93] G. J. Sullivan. Multi-Hypothesis Motion Compensation for Low Bit-Rate Video Coding. In *Proceedings of the IEEE International Conference on Acoustics, Speech and Signal Processing*, volume 5, pages 437–440, Minneapolis, MN, USA, April 1993.

[SW98] G. J. Sullivan and T. Wiegand. Rate-Distortion Optimization for Video Compression. *IEEE Signal Processing Magazine*, 15(6):74–90, November 1998.

[TH81] R. Y. Tsai and T. S. Huang. Estimating Three-Dimensional Motion Parameters of a Rigid Planar Patch. *IEEE Transactions on Acoustics, Speech and Signal Processing*, 29(6):1147–1152, December 1981.

[TKI97] Y. Tomita, T. Kimura, and T. Ichikawa. Error Resilient Modified Inter-Frame Coding System for Limited Reference Picture Memories. In *Proceedings of the Picture Coding Symposium*, pages 743–748, Berlin, Germany, September 1997.

[Uns99] M. Unser. Splines: A Perfect Fit for Signal and Image Processing. *IEEE Signal Processing Magazine*, 16(6):22–38, November 1999.

[WA94] J. Y. A. Wang and E. H. Adelson. Representing Moving Images with Layers. *IEEE Transactions on Image Processing*, 3(5):625–638, September 1994.

[Wan95] B. A. Wandell. *Foundations of Vision*. Sinauer Associates, Inc., Sunderland, MA, USA, 1995.

[WB95] P. Wong and D. Britland. *Mobile Data Communications Systems*. Artech House, Norwood, MA, USA, 1995.

[Wed99] T. Wedi. A Time-Recursive Interpolation Filter for Motion Compensated Prediction Considering Aliasing. In *Proceedings of the IEEE International Conference on Image Processing*, volume 3, pages 672–675, Kobe, Japan, October 1999.

[WFG98] T. Wiegand, M. Flierl, and B. Girod. Entropy-Constrained Linear Vector Prediction for Motion-Compensated Video Coding. In *Proceedings of the IEEE International Symposium on Information Theory*, page 409, Boston, USA, August 1998.

[WFSG00] T. Wiegand, N. Färber, K. Stuhlmüller, and B. Girod. Error-Resilient Video Transmission Using Long-Term Memory Motion-Compensated Prediction. *IEEE Journal on Selected Areas in Communications*, 18(6), June 2000.

[Wic95] S. B. Wicker. *Error Control Systems*. Prentice Hall, Englewood Cliff, NJ, USA, 1995.

[WLCM95] T. Wiegand, M. Lightstone, T. G. Campbell, and S. K. Mitra. Efficient Mode Selection for Block-Based Motion Compensated Video Coding. In *Proceedings of the IEEE International Conference on Processing*, volume 2, pages 559–562, Washington, D.C., USA, October 1995.

[WLG98] T. Wiegand, B. Lincoln, and B. Girod. Fast Search for Long-Term Memory Motion-Compensated Prediction. In *Proceedings of the IEEE International Conference on Processing*, volume 3, pages 619–622, Chicago, USA, October 1998.

[WLM+96] T. Wiegand, M. Lightstone, D. Mukherjee, T. G. Campbell, and S. K. Mitra. Rate-Distortion Optimized Mode Selection for Very Low Bit Rate Video Coding and the Emerging H.263 Standard. *IEEE Transactions on Circuits and Systems for Video Technology*, 6(2):182–190, April 1996.

[WLV98] J. Wen, M. Luttrell, and J. Villasenor. Trellis-Based R-D Optimal Quantization in H.263+. *IEEE Transactions on Circuits and Systems for Video Technology*, 1998. Submitted for publication.

[WS91] H. Watanabe and S. Singhal. Windowed Motion Compensation. In *Proceedings of the SPIE Conference on Visual Communications and Image Processing*, volume 1605, pages 582–589, Arlington, TX, USA, 1991.

[WZG99] T. Wiegand, X. Zhang, and B. Girod. Long-Term Memory Motion-Compensated Prediction. *IEEE Transactions on Circuits and Systems for Video Technology*, 9(1):70–84, February 1999.

[YMO95] Y. Yokoyama, Y. Miyamoto, and M. Ohta. Very Low Bit Rate Video Coding Using Arbitrarily Shaped Region-Based Motion Compensation. *IEEE Transactions on Circuits and Systems for Video Technology*, 5(6):500–507, December 1995.

[Yua93] X. Yuan. Hierarchical Uncovered Background Prediction in a Low Bit-Rate Video Coder. In *Proceedings of the Picture Coding Symposium*, page 12.1, Lausanne, Switzerland, March 1993.

[ZBK97] K. Zhang, M. Bober, and J. Kittler. Image Sequence Coding Using Multiple-Level Segmentation and Affine Motion Estimation. *IEEE Journal on Selected Areas in Communications*, 15(9):1704–1713, December 1997.

[ZK98] K. Zhang and J. Kittler. A Background Memory Update Scheme for H.263 Video Codec. In *Proceedings of the European Signal Processing Conference*, volume 4, pages 2101–2104, Island of Rhodes, Greece, September 1998.

[ZK99a] K. Zhang and J. Kittler. Using Scene-Change Detection and Multiple-Thread Background Memory for Efficient Video Coding. *Electronic Letters*, 35(4):290–291, February 1999.

[ZK99b] Q. F. Zhu and L. Kerofsky. Joint Source Coding, Transport Processing, and Error Concealment for H.323-Based Packet Video. In *Proceedings of the SPIE Conference on Visual Communications and Image Processing*, volume 3653, pages 52–62, San Jose, CA, USA, January 1999.

[ZRR00] R. Zhang, S. L. Regunathan, and K. Rose. Video Coding with Optimal Inter/Intra Mode Switching for Packet Loss Resilience. *IEEE Journal on Selected Areas in Communications*, 18(6), June 2000.

Index